MW01014751

Harvesting Urban Timber

Vincennes University
Jasper Campus Library
900 College Ave.
Jasper, IN 47546

Harvesting Urban Timber

A Complete Guide

Sam Sherrill

LINDEN PUBLISHING
FRESNO, CA

Harvesting Urban Timber

Text © 2003 by Samuel B. Sherrill
Computer-generated graphics © 2003 by Carey William Sherrill
Cover artwork by Carey William Sherrill

All rights reserved. No part of this book may be reproduced or transmitted in any form or by any means, electronic or mechanical, including photocopying, recording, or by any information storage and retrieval system, without written permission from the publisher, except for the inclusion of brief quotations in reviews.

ISBN 0-941936-71-6
135798642
First printing: July 2003
Printed in Singapore

Library of Congress Cataloging-in-Publication Data
Sherrill, Samuel B.
Harvesting urban timber : a complete guide / Samuel B. Sherrill
p. cm. -- (Woodworker's library)
Includes bibliographical references and index.
ISBN 0-941936-71-6 (pbk. : alk. paper)
1. Trees in cities--Utilization. 2. Tree felling. 3. Wood waste--Recycling. I. Title. II. Woodworker's library (Fresno, Calif.)
SB436.S52 2003
363.72'88--dc21

2003003973

Your safety is your responsibility. Neither the author nor the publisher assume any responsibility for any injuries suffered or for damages or other losses incurred that may result from material presented in this publication.

LINDEN PUBLISHING

Linden Publishing Inc.
2006 S. Mary St.
Fresno, CA 93721 USA
tel 800-345-4447
www.lindenpub.com

Contents

Acknowledgments

In writing this book, I am very fortunate to have had the encouragement and support of family, friends, Wood-Mizer Products, Inc., and the USDA Forest Service. They and others all gave generously of their time and effort. The book is much better for their contributions. Whatever shortcomings and errors remain are far fewer in number than would have otherwise been the case.

I do not believe I have encountered a more dedicated group of public servants than those who work in the field of urban forestry in the USDA Forest Service. In my judgment, taxpayers are really getting their money's worth from the work these people do for all of us who live in urban America. In particular, for their invaluable help, I would like to thank Steve Bratkovich, David Nowak, Bill Simpson, Gregory McPherson, Ed Cesa, and Kevin Smith. John "Rusty" Dramm carefully read and commented in detail on the section about sawing logs for high quality lumber: no other part of the book was so much improved by any one person's review than this one.

Eric Oldar of Urban and Community Forestry, California Department of Forestry and Fire Protection, reviewed the entire book and has provided both encouragement and insight from the beginning. In my view, California is at the cutting edge of making green waste into a green resource because of Eric's tireless efforts. In turn, urban foresters in California, such as Cindy McCall of Lompoc, have been able to start innovative and cost-reducing projects in their own cities.

Barb Henderson, Executive Director of the Forest Industry Safety and Training Alliance (FISTA), Inc. carefully reviewed the sections on safe practices of tree felling and chainsaw use. Safety is a very important issue that must be foremost in the minds of those who want to do their own felling and log cutting. Because Barb and FISTA know what they are talking about, I did not deviate from their recommendations.

For the past seven years, I have received continuous support and encouragement from Wood-Mizer Products, Inc., the largest manufacturer of portable band mills. They donated a small mill to the harvesting urban timber effort, printed a booklet describing the basics of harvesting urban timber, several times loaned me a mill to practice on, cheerfully conducted mill demonstrations every time I asked, kept me up to date on new machines and technology, and reviewed parts of the book. In my experience, this is a fine company that treats its employees well and makes a variety of high-quality band mills that can be put to good use in urban areas throughout the world. In particular, I want to thank Jim "JB" Brown, John Hicks, Bob Hooten, Susie Rudd, and Dave Mann.

Russ Morash brought *The New Yankee Workshop* to Cincinnati several years ago to feature the idea of making the best use of discarded urban trees. Norm Abram made a sink base from cherry salvaged from a tree that otherwise would have ended up in a local landfill. This one program has encouraged an ever-growing number of woodworkers to think more about making use of wood that otherwise would have been thrown away or burned. I am indebted to Russ and Norm for their support and to Russ for reviewing the entire book.

Steve Shanesy, editor of *Popular Woodworking* Magazine, deserves credit for first having the foresight to bring the idea of reclaiming urban trees to the attention of the readers of woodworking magazines. His effort continues to pay off. I am still receiving calls and mail from readers all over the country who first learned about this in *Popular Woodworking*.

Several others deserve recognition for helping to pave the way for a book of this kind. Ed Cesa, of the USDA Forest Service, and Ed Lempicki, Chief of the New Jersey Forest Service, wrote *Recycling Municipal Trees* (published by the USDA Forest Service) in 1994 based on their own efforts very early in the last decade. Steve Bratkovich's *Utilizing Municipal Trees: Ideas From Across the Country* (also published by the USDA Forest Service) brings our attention to successful local efforts across the country.

I am grateful to my close friends, Carolyn and Lowell McCoy, who reviewed and commented on the entire book. Their encouragement, questions, and comments sharpened my thinking and my writing. Jack Butcher, President of Madison Tree Service, reviewed the section on chain saw safety and tree felling. He also gave me logs, and space to park and use the band mill. Thanks also to Jeff Stromatt for helping to promote the idea. Ed Motz has sawn logs for me for almost ten years. I learned much by watching and talking with him.

Finally, I am most indebted to my family who read, commented, and encouraged. My brother Barry read the entire manuscript, as did my son Carey and my wife, Pat. Carey designed the cover and created all of the computer-generated images of logs. He also designed and maintains the HUT website,

www.HarvestingUrbanTimber.com.

And, Carey and Pat were my partners in the various projects that comprise most of the experience behind this book. Most important of all, Pat pushed logs, lifted boards, cheerfully endured many wood-related outings, and numerous times asked the questions I missed and remembered points I forgot to write down. In the most meaningful way, as my partner in this effort, she is also the silent co-author of my book and my life. Without her, neither would be complete.

Sam Sherrill
Milford, Ohio
April, 2003

Introduction

The Downhill Racer

Until I took wood shop in high school, my only contact with wood was climbing trees. I discovered in shop how much fun it is to work with wood. I was so impressed that the following summer I tackled my first project: a soapbox racer made from two-by-fours and plywood. The racer combined my newfound enthusiasm for woodworking with a teenage boy's interest in anything to do with cars.

At seven feet long and almost three feet wide, what I built that summer was the road yacht of soapbox racers. The steering mechanism was a rope wrapped around a broom handle, with a steering wheel fastened at one end. The other end was held in place by a two-by-four with a hole drilled through it. Steering was not very responsive because the rope stretched under tension. Several full rotations of the steering wheel turned the front wheels a few degrees. The brake was a wooden lever that rotated on a bolt. When I pushed on one end of the lever the other end pressed on the right back wheel, which, by my theory, would bring the racer to a halt.

On the initial, and as it turned out, final run, getting started was the first challenge. Once underway, stopping was the second and greater challenge. At first, I could not even get the racer to roll down the street in front of my house. It sat stubbornly at the crest of the hill, seemingly just out of the pull of gravity, until my mother gave it a good nudge with the family Buick Roadmaster. Then it began to roll, and then roll faster and even faster. As it gained speed, I began applying the brake, with little effect. Finally, the bolt came loose and the brake lever fell off. I remember careening down the street with no brakes and not much steering. This had not been covered in wood shop, or any other class I had taken. From physics, I recalled something about accelerating at thirty-two feet/second/second, but that was about going faster, not slower. Finally in desperation I managed to steer the racer over the curb and through several front yards until, appropriately enough, it collided with a tree. Fortunately, I and the racer and the tree all survived the impact, as did my enthusiasm for making things out of wood. I have been a woodworker ever since, but now my projects sit firmly in one place: pushing, steering, and braking are not required.

Urban Sources of Wood

From the summer of the soapbox racer until the mid-90s, I bought the wood I needed from lumberyards and home centers. I never gave much thought to where it came from or how it ended up in retail outlets. I shopped for lumber like I shopped for groceries, automobiles, or any other consumer goods, by looking for the best quality at the lowest price.

I first learned how trees are transformed into the lumber when I visited a cabinetmaker in rural southern Indiana. He bought trees from local farmers, had the logs milled and dried locally, and used the lumber in the cabinets and furniture he made for his clients. Over several visits, he walked me through the basic process, from the forest to the mill to the shop. For the first time, I began to understand and appreciate just what is involved in manufacturing lumber.

Not long after that, I was asked if a downed tree in a local city park might be usable. That question, plus what I had learned from the cabinetmaker, sparked my interest in the possibility of using local urban trees as a source of lumber. Steve Shanesy, editor of *Popular Woodworking*, featured the idea in several issues of the magazine beginning in November, 1997, raising the possibility that this was more than a local issue.

That this idea is feasible for other cities was confirmed in my mind by a visit to Cincinnati in late 1998 from Norm Abram and Russ Morash, producer and director of *The New Yankee Workshop*, a PBS television program. I had contacted Russ earlier that year to see if this local effort was something that might interest viewers of the program. He was very interested and agreed to come to Cincinnati to film the felling and sawing of a street tree as an introductory segment for a show that used cherry lumber sawed from a local yard tree I supplied. At my request, a Cincinnati Parks Department crew timed the removal of a large dying red oak, situated between the sidewalk and the street in a Cincinnati neighborhood, to coincide with the visit from Norm, Russ, and the New Yankee crew. Also at my request, Wood-Mizer Products, Inc., a manufacturer of portable band mills, towed a mill to Cincinnati to saw the oak logs into lumber. As pictured at right, Norm and Russ and the crew filmed the tree being felled and sawed into lumber. For the viewers of the program, this was a direct demonstration in an urban setting of how a street tree could be converted into something far more useful and valuable than firewood or mulch (or worse, be discarded as trash). To this day, I can tell what part of the country that particular episode of *The New Yankee Workshop* is playing on public television by where phone calls, letters, and email come from shortly thereafter.

I'm still exploring the possibilities that have arisen from these and related experiences. This book contains what I have learned so far.

What is an Urban Tree?

While few of us could precisely define a tree, most of us know one when we see it. Here, and again in the next chapter, a tree is defined as a woody perennial plant with either one or just a few main stems. At full growth, a tree is generally larger than other plants such as shrubs (which are also woody perennials but with many stems).

Understanding whether a woody perennial is a tree is easier than understanding exactly what is meant by "urban". There are

FIGURE 1
Norm Abram and the author standing next to a large red oak that has just been felled by a crew from the Cincinnati Parks Department.
Photo by Patricia Sherrill

FIGURE 2
Rain is no deterrent. Norm takes a turn on the mill while author watches and carries the umbrella.
Photo by Patricia Sherrill

two different but overlapping geographic definitions used by the federal government to count urban populations: one is based on places identified by names and accepted boundaries (such as the corporate limits of a town or city) while with a few exceptions the other is more broadly based on counties.

For the 1990 census, the U.S. Bureau of the Census defines "urbanized areas" as places with a minimum population of 50,000 and a population density of 1,000 residents per square mile. "Urban places" must have a minimum population of 2,500, and must be outside urbanized areas.[1] Places with fewer than 2,500 residents outside urbanized areas are considered rural. In Georgia, for example, Atlanta (population 416,474) is an urbanized area while Valdosta (population 43,724) is an urban place. Plains, Georgia (population 637) is rural.

FIGURE 3
A French side table made by Norm Abram from black walnut I acquired from a local tree service company just days before the log would have been cut into firewood.
Photo by Keller & Keller for WGBH

FIGURE 4
Some of the lumber from the red oak tree was used by my son, Carey, to make this workbench for my shop.
Photo by Carey Sherrill

In general, the Office of Management and Budget (OMB) in the Executive Branch defines major urban populations and centers by counties, not places.[2] By OMB's definition, metropolitan areas have one or more large core populations that are socially and economically linked to surrounding and nearby counties.[3]

The 20-county Atlanta metropolitan area's population was almost 4 million in the year 2000. The nation's largest metropolitan areas are comprised of several large population centers and adjacent counties all linked together. With a population of more than 21 million, the largest is the New York metro area that covers 36 counties in New York, Connecticut, New Jersey, and Pennsylvania. Most but not all urbanized areas and urban places are in the nation's metropolitan areas.

Using the county-based definition of metropolitan areas, about eight out of ten of us are urban residents. Whether we live in hamlets, villages, small towns, cities, or in one of the largest metro areas, such as New York, the trees that grow in these areas are urban trees. By the Forest Service's estimate for the contiguous 48 states, we are talking about 74 billion trees, an amount equal to about one-fourth of the estimated 319 billion live trees growing in the nation's commercial timberland.[4]

We think of urban trees as one of the essential parts of the urban forest. However, the urban forest also consists of soil, water, surrounding vegetation (shrubs and grass, for example), wildlife inhabitants, human inhabitants, and human structures from buildings and parking lots to above- and below-ground utilities and transportation systems. All of these components are interspersed among and interrelated with one another.

Thinking differently about the trees that eighty percent of us see every day is, in general, what this book is about. More specifically, what we can do to extract greater value from the trees that come down every year is what I want to bring to your attention. Though this book focuses on specific ways of making better use of urban trees, I do not want us to lose sight of the nature and separate importance of the larger urban forest of which the trees are an essential part. To paraphrase an old cliché, we must not lose sight of the forest even as we focus on the trees themselves. What we do, or do not do, with urban trees will have a significant effect on the urban forest and ultimately on the quality of urban life.

Lessons Learned from Others and from Experience

The content of this book is what I have learned from my own experience (including mistakes), research on wood processing, and from discussions with experts in urban forestry and urban wood processing. Several lessons stand out from what I have learned so far.

Where it Grows Determines Where it Goes

I learned that what a downed tree becomes depends less on the kind of tree it is than on the place where it grows. For example, cherry trees from the forests of central Pennsylvania are likely to end up as fine furniture or cabinetry. Comparable trees from the nation's urban areas will end up as mulch, firewood, or garbage (identified by the milder technical term "green waste" which is part of the much larger municipal solid waste stream). On some construction sites, trees felled to make room for buildings are buried on-site. In more than a few cases, they are simply burned. I am struck by the irony of how much we value urban trees as long as they are standing and alive but, as soon as they hit the ground, they lose all value and become costly garbage instead.

Not All Urban Wood Can Become Lumber

I learned that not all green waste could be converted into usable lumber. As a rule tree limbs, even large ones, do not make good lumber because of compression and tension stresses created by the way they grow (although I describe a major exception from my own experience in Chapter Four). Small or short trunks and large hollow or otherwise severely damaged trunks yield little if any usable lumber. Even large solid trunks with some usable wood in them might not be worth the effort if they are filled with concrete, riddled with metal, or have grown under stressful conditions. Of course, bark and leaves are of no use either. At most, roughly one-third of all of the nation's green waste is usable as lumber. The remaining two-thirds is usable only as mulch or firewood. Even so, by one estimate, approximately 3 to 4 billion board feet of potential lumber are thrown away every year as green waste.[5] (The board foot is the standard volumetric measure of lumber. It is equal to a board that is 1 inch thick and 12 inches square, or 144 cubic inches.)

To put this in a meaningful context, the consumption of hardwood in the U.S. in 1999 was 14 billion board feet.[6] The annual amount of hardwood lumber we throw away is equal to about one-fourth of what we consume.

But, two critical questions must be asked about this 3 to 4 billion feet of wood: can it actually be made into lumber? As explained in Chapters One through Three, the technical answer is that it can. The

FIGURE 5
Urban trees felled for development can become logs or mulch. Better yet, the best of the trees can become logs and the rest mulch.
Photo by Sam Sherrill

schools, vocational centers, and craft centers. For cities, utilizing urban timber can reduce the amount of green waste going into landfills, many of which are reaching their limits.

4. Both experienced and first-time sawyers will find useful suggestions on establishing sources of saw logs and potential markets for lumber. A sawyer might, for example, establish a working relationship with the local park board to cut and share lumber cut from park trees. A similar arrangement can be made with commercial tree services and local utility companies. In cooperation with local woodworkers, sawyers could mill logs into lumber that could then be made into furniture, or into trim for the homes and offices that displaced the trees.
5. Owners and operators of urban tree-service companies will find reasons to think about expanding their businesses to include sawing logs into lumber. Some larger companies have purchased existing sawmills as a way of reducing or eliminating their disposal fees while adding to the revenue and profit. Others could begin more modestly by purchasing a transportable sawmill or even arranging with a sawyer to cut and share. These options are even more important to companies operating in states that restrict or ban the dumping of green waste.
6. If you are a volunteer who contributes time to zoos, parks, or nature organizations, this book describes another way you can serve your community by knowledgeably promoting the use of urban timber, especially for public projects.
7. Finally, even if you are not directly involved with or interested in wood, you can nevertheless appreciate ideas and stories about others who have found innovative ways to use a resource we all know and for which we have a deep affection.

Content

Briefly, *Harvesting Urban Timber* is organized as follows: Chapter One describes how three states are managing the problem of green waste by promoting the value and uses of urban trees as sources of lumber. Chapters Two and Three describe how trees can be safely felled, cut into logs, sawed into lumber, and dried. Chapter Four describes eight different urban timber projects, starting with my own. Chapter Five outlines how individuals can start harvesting urban timber efforts in their own communities.

Additional details on subjects covered in the book are in the Endnotes, as are sources and additional references. Details and references are provided for readers who are not familiar with the subject and need additional information, as well as for those who want to delve into the subject more deeply than the material covered in the body of the book. Some of the notes are lengthy but, of course, these and all the rest are optional reading.

1 FROM WOOD WASTE TO WOOD RESOURCE

Introduction

From the time our earliest ancestors first learned to use the raw materials at hand, they used wood from their immediate surroundings for food, fuel, weapons, and eventually to build and furnish their shelters. As settlements became towns and cities, and trees were cut to make way for growing population centers and for farming, wood sources shifted to more distant forests where they have since remained. Today we are again looking at the wood that comes from our immediate urban surroundings, not because it is necessarily in short supply, but because urban trees have become part of this nation's urban waste disposal problem. Following the example set by our ancestors, perhaps we can learn once again to make better use of the wood close at hand.

Green Waste, or a Green Resource Being Wasted

The trees in our yards, on our streets, and in our city parks are amenities as long as they are standing and alive. As soon as they come down, they are treated as something to be thrown away. As garbage, or green waste, downed urban trees enter the municipal solid waste stream, a potpourri of stuff that ranges from worn-out appliances to food scraps, clothing, boxes, auto tires, shipping pallets, and paper. Excluded by definition from the waste stream—but nevertheless thrown away in urban areas—are automobile bodies, construction and demolition debris, industrial wastes, sludge, and ashes. A mass consumption society such as ours is also a mass disposal society. According to the BioCycle, in 1999, residents of American cities threw away almost 400 million tons of solid waste.[1] However, according to the EPA's latest estimate, which is for 1997, 217 million tons of solid waste was generated.[2] The BioCycle estimate is larger because, by virtue of the way they collect the numbers from a survey of states, their definition of solid waste is broader than the EPA's. Also, it is two years more recent than the EPA's estimate. Regardless of which definition we use, or exactly when the amount is estimated, this is a lot of garbage.

Neither the BioCycle nor the EPA data are adequate for estimating the amount of potential saw logs discarded annually into the municipal waste stream. The NEOS Corporation estimated for 1994 that of the 200 million cubic yards of green waste generated, about 15 percent, or 30 million cubic yards, was in the form of unchipped logs more than one foot in diameter.[3] Converting this number to sawn board feet gives us the 3 to 4 billion board feet of potential lumber cited in the Introduction. I believe this could be an underestimate, since few trees are being cut into saw logs. Instead, they go uncounted when they are either disposed of on-site (buried or burned) and undercounted when cut into small sections for easier hauling and dumping. Either way, these are rather crude estimates, but the best we have because this issue has only recently begun receiving attention from government and from organizations that might collect more precise data.

Regardless of what the exact numbers are, we know that yard, ground, and urban tree maintenance and land development generate a substantial amount of waste that traditionally has been buried, burned, deposited in landfills, cut up for firewood, or ground up for mulch. Large-scale burning is no longer widely used because it pollutes the air in our cities. Landfills serving urban areas are filling up and decreasing in number. In 1999 there were 2,216, or almost 100 fewer than than the 2,314 operating in 1998.[4]

Siting the occasional new urban landfill anywhere near residential areas is always a contentious and controversial issue. Through legislation and local ordinances, state and local governments are placing much greater emphasis on reducing the size of the waste stream at its sources. About half the states currently ban green waste from landfills. On their own, many cities and towns are following the same policy. So are local parks departments: for example, the Cincinnati Park Department's policy is not to dump green waste into landfills, nor is it burned. In addition to source reduction, legislation requires more reliance on the reclamation, recycling, and reuse of what remains after reduction. Whatever cannot be reduced at the source can be used before or as it reaches landfills. If we treat urban trees as a source of lumber, not garbage, or even just firewood and mulch, we can reduce green waste at its source and make better use of what remains.

In my view, all green material currently being discarded in urban areas could be used as a source for firewood, mulch, lumber, and co-generation fuel for urban power plants. I can foresee a day in the not-too-distant future when all green materials, from trees to shrubs, from leaves to grass, would be put to their highest and best use and nothing would be wasted. But that would be the subject of another book.

In this chapter, I describe recent efforts in three states, New Jersey, Minnesota, and California, to encourage local governments and private businesses to find greater economic value in urban trees. I picked these states because, so far as I can determine, their waste reduction policies have included specific efforts to convert trees into logs and lumber. Although their policies differ somewhat, they all have in common the primary task of adjusting attitudes about urban trees once they have been felled. Also, thinking about the economic value of urban trees leads us to think more about what, in the long run, we want from our urban forests and how we plan their futures.

The New Jersey Municipal Log Project

The Expense of Dumping Good Wood

In the late 1980s, Ed Lempicki of New Jersey's Forestry Services (NJFS) found himself responding to a growing number of calls about the disposal of urban trees.[5] Both municipal officials and private tree services around the state expressed concern about the number of trees being felled, about the high quality of some of the logs being thrown away, and about rising disposal costs. There was also concern that local landfills could not continue to accept the rising tide of green waste. This was an issue that required examination and possible action at the state level.

Through NJFS, Ed applied for and, in 1991, received a grant from the USDA Forest Service's Rural Development Through Forestry program. The grant financed a survey among the state's 567 local governments to determine the magnitude of the tree disposal problem. The results confirmed what Ed had tentatively surmised from the calls. Good saw logs were being thrown away and disposal costs were rising. The costs were becoming a significant burden for local governments, reaching about half of some municipal shade-tree-management budgets. Local governments and tree services were looking to the state for help. The federal grant enabled Ed and NJFS to document the problem and to provide direct technical assistance to participating communities.

Assistance focused on treating street trees as a source of marketable saw logs. We are all aware that throughout their lives, trees are valuable in important ways: aesthetic appeal, erosion control, energy conservation, wildlife habitat, and air cleansing capacity. New Jersey was the first state to formally act on the idea that urban trees could have value after they have fallen. This was a new kind of logging.

Urban Versus Commercial Logging

In several important ways, felling trees in urban areas is the opposite of commercial logging in the nation's forests. In urban areas, we have to wait until natural forces such as wind, disease, infestation, or age bring down trees; because they are public hazards or are damaging hardscape (sidewalks, for example) or are felled when land is cleared for development. In commercial logging, trees are identified and harvested based on a range of considerations such as species, size, and quality. Or, they are harvested from tree farms where specific species, such as long leaf pine, have been grown strictly for commercial pulp and lumber markets. By contrast, urban logging requires us to wait for the trees to come down for other reasons and then make use of what has become available.

This is a more challenging form of logging: we do not get to pick what we want, we must be prepared to want what has been already picked for other reasons and by other forces. This was the challenge Ed faced: how to add value to street trees in New Jersey being felled in relatively small quantities, at irregular intervals, and for reasons unrelated to species, size, and quality. Working with Ed Cesa, a USDA Forest Service marketing specialist, Ed Lempicki developed a set of easy-to-follow principles for identifying marketable urban logs by size, species, and quality (including awareness of foreign material embedded in the wood).[6] Information was provided on proper log manufacturing (cutting to commercial lengths, for example), storage, and loading for transportation. They also identified ways to find log and lumber buyers, from sawmills to amateur woodworkers. Safe procedures for cutting and loading were identified as well.

They published the principles and information they used in the New Jersey project in *Recycling Municipal Trees: A Guide to Marketing Sawlogs from Street Tree Removals in Municipalities.*[6]

In addition to describing how to identify, cut, and market logs, *Recycling* also describes how a tree-service company and a sawmill successfully processed street trees. *Recycling* is an important publication. In addition to being widely distributed (at no charge), it was selected in 1995 for the USDA Forest Service's Technology Transfer Award.

Results of the New Jersey Project

Of the 200 municipalities that expressed initial interest in the project, only about 20, or 10 percent, established operating programs. About 25 percent received some form of assistance.[7] As long as the grant ran, NJFS provided participating cities and towns with the technical support needed to identify and sell logs.

Like many federal grants, the purpose of this one-year grant was to test an idea that—if it worked—would be ongoing in New Jersey and replicated across the nation. As far as it went in one year, the New Jersey effort seemed to work. That the project did not continue at the same level of effort after the first year was a loss for the state. However, the much greater cost to the nation's cities and towns is the lost opportunity to have learned of and started urban timber reclamation programs of their own.

As a rough estimate, more than 30 billion board feet of lumber were disposed of as green waste during the 1990s. About 3 billion board feet of lumber would have been produced if enough cities and towns had started programs, during that time, to reclaim just 10 percent of the 30 billion board feet thrown away. Assuming a market value of about 75 cents per board foot, the 3 billion board feet would have been worth roughly $2 billion. To the $2 billion in lost lumber, we must add green waste disposal costs, including landfill fees, and the net cost of lumber purchased for park and other public projects that, at least in part, could have been constructed with much less expensive reclaimed wood. The marginal cost of converting trees to lumber would be very small since the disposal effort, and cost, which would have been expended anyway, is redirected to reclamation instead of disposal. In effect, sawmill time is substituted for chainsaw time. The additional time and expense of cutting lumber reduces the amount of time and expense spent cutting tree trunks into fireplace lengths (there is a safety improvement as well since mills are safer to operate than chain saws). In short, if about the same amount of money is going to be spent either way then, whenever possible, it should be spent in the way that yields the most valuable mix of wood products. Output would be a blend of various qualities of lumber as well as firewood and mulch. Little or none would be costly garbage.

A Mill of Their Own

Following the example it set for others, in 1999 the NJFS bought its own portable sawmill (Figure 6) and now cuts the lumber it needs from urban trees. The mill also provides a unique educational experience

FIGURE 6
The new New Jersey Wood-Mizer® sawmill, and admirers.

for students. Kids in New Jersey now have the opportunity to learn where the wood in their own lives comes from and how it is produced. For them, wood becomes something more than just another consumer product purchased at lumberyards and home centers. Ed Lempicki continues to provide support and information for those who call, and there are still calls for both.

Integration of Tree Service and Sawmill Operations

The New Jersey project may be the catalyst for horizontal mergers of tree services and sawmill companies. Ed said, "A number of our standard sawmills have been bought out by tree removal businesses." He pointed out that the mergers "...allow the tree-service companies to capture the valued-added opportunities that come with owning sawmills". Further, he noted that sawmill ownership also provides a way for tree-service companies to reduce their disposal costs, often a large part of their total operating costs.[9,10]

We would expect this trend toward horizontal integration to be even more pronounced in states charging high tippage fees at landfills. The trend would be most pronounced in states that impose limits on green waste dumping or ban it outright. Smaller tree-service companies unable to purchase sawmills can extend their operation by purchasing a transportable mill. Compared to large stationary sawmill operations, portable mills are less expensive to purchase and easier to operate.

The New Jersey project was the first organized governmental effort to identify downed urban trees as a source of lumber, not just as firewood and mulch. From the larger perspective of the recycling movement, it was the first to find a practical way for both public agencies and private businesses to reduce green waste. As an example and a source of encouragement, the New Jersey Municipal Log Project continues well beyond its one year of actual operation.

Minnesota Urban Tree Residue Utilization Project

Dutch Elm Disease

Dutch elm disease is caused by a fungus transported by beetles that burrow under the bark of the tree. The fungus spreads inside the elm's sapwood, the relatively narrow band of wood just beneath the bark that carries water and nutrients from roots to leaves. Much as cholesterol clogs the blood vessels in our own bodies, the fungus blocks the sapwood vessels of the elm, preventing the flow of water and nutrients essential to a tree's life. Leaves turn yellow and branches fall off as the fungus spreads. Eventually, the entire tree succumbs.[11, 12] Over a forty-year period, the disease killed about 100 million American elms. The Elm Research Institute, organized in 1967, developed an effective fungicide that is saving the surviving trees. The Institute also developed a disease-resistant tree, the liberty elm, which has been available since 1983 to communities through the country on a cost-share basis. So far, about one-quarter million liberty elms have been planted throughout the U.S.[13]

Until the disease began spreading in the 1930s, elm was widespread throughout the cities and towns of the eastern and midwestern United States, including Minnesota. Minneapolis and St. Paul were especially hard hit during the 1970s when elms began dying off in massive quantities. The state responded by providing funds for disposal by burning and burial in landfills. Some attempts were made to use the wood but without much success. After the problem of disposing of the dead elms diminished so did interest in and funding for alternatives to burning and burying. However, in terms of disposal, the state and the cities had gotten a glimpse of a future that was closer than they might have realized at the time.

Filling Minnesota Landfills

In 1982, the Twin Cities relied on eleven sanitary landfills. By 1989, there were five left and, at that time, they were expected to fill up in about ten years.[14] State and local governments were compelled to look again at the problem of disposal. Now, waste management had to be viewed, not as a short-term problem posed by the disposal of dead elm trees, but on a more comprehensive and long-term basis.

In 1989, the Minnesota legislature passed the Waste Management Act. This legislation set the rules for how county governments throughout Minnesota would manage their municipal solid waste stream. Counties are required to create waste disposal plans that emphasize reuse over dumping. It also set specific goals for reducing the volume and the toxicity of the stream that flowed into the state's landfills. Recycling, reuse, and outright prohibition are identified as ways to reduce volume and toxicity. Disposing of waste in landfills is the least preferred way, the last resort.

The Act specifically bans the dumping of yard waste into the landfills (yard waste is the grass clippings, garden debris, leaves, and prunings we generate in the upkeep of our lawns and commercial green spaces). Urban tree residue—logs, stumps, limbs, leaves, and chips produced by grinders—is not defined as yard waste by the Act. Instead, it is identified as a potential source of energy (firewood) soil additives (mulch) and, where possible, as a source of lumber.

A Project to Reduce Urban Green Waste

With grants from the USDA Forest Service, the Minnesota Department of Natural Resources (MDNR) and Minnesota Shade Tree Advisory Committee jointly explored ways to use urban tree residue in the Twin Cities area. One of the most fundamental ways to change traditional views, and ways of doing things, is to gently persuade people to simply change their language. Words not only carry specific meanings but also larger connotations, often pejorative, which shape how we feel and what we do as a result. The joint effort between the Department and the Committee, carried out by a Task Force, focused on redefining urban tree residue as a potential resource; something to be used, not just thrown away.

Additional objectives were finding markets for urban tree products, persuading Twin Cities decision-makers to participate in the search for new ways to use the trees, and collecting and distributing information to all involved (homeowners, tree services, local government, and community organizations) about how this effort could work with their informed participation.

This project ran from mid-1991 through the end of 1994. A joint report, *Final Report: Urban Tree Utilization Project,* was issued in June, 1994.

At the outset, the Task Force commissioned a survey to determine how much residue was being generated in the Twin Cities area and what was happening to it.[15]

The survey revealed that early in the decade of the 1990s, the metro area produced a total of about 326,000 tons of tree residue annually. About half came from tree services, the other half from utility companies, public agencies, and land cleared for development. Although over half of the total residue was chipped, about one-fourth emerged as logs. By weight, that fourth amounted to about 85,000 tons. The producers themselves cut only 4,000 tons, or about 5 percent, of the logs into lumber. They sold about 15,000 tons, gave away almost 55,000 tons, temporarily stored just under 2,000 tons and either chipped or burned the remaining 9,000 tons.

Unfortunately, there is no way of knowing what became of the 70,000 tons of logs that were sold or given away. While most may have been cut up for firewood, at least some of it may have been milled into lumber.

Two commercial outlets for logs were pursued. The MDNR assisted Henderson Hardwoods in developing a facility for producing lumber that would be us7ed to repair wooden pallets, an ideal use for low-grade lumber. Unfortunately, this particular commercial venture was not fully tested because, for reasons unrelated to the project, Henderson went out of business. Phil Vieth, now retired from the MDNR, worked closely with the Henderson Company. He told me he believes that the company, with a stronger financial structure, would have succeeded.[16]

A second outlet was Minnesota Hardwoods, Inc., a lumber company in Courtland, Minnesota that operates its own sawmills and sells the lumber it saws. Phil told me that this company started buying urban logs even before Henderson, and is still occasionally buying them around the Twin Cities area. Urban foresters notify the company after they have selected street trees for removal. Minnesota Hardwoods would then buy the trees from the service that felled them, mill the logs, and sell the lumber. Frank Kilibarda, owner of Minnesota Hardwoods, told me that urban trees are more difficult to cut because of the metal often embedded in them.[17] His company uses metal detectors and circular saw blades capable of cutting through most metal. Even so, he feels that the logs are worth the effort.

Phil Vieth came to the Twin Cities in the mid-1970s to work as a forest products utilization and marketing specialist for MDNR. He worked directly on the elm tree disposal problem and on subsequent projects that focused on finding new uses for tree waste. I asked Phil whether he thought the specific effort to mill lumber from Twin Cities trees had been successful. He said he thought that it "had been very successful", even though there is not as wide a variety of high quality hardwoods growing in the Twin Cities area as there is, for example, in New Jersey or Ohio.[18]

Attitude Adjustments About Urban Trees

Phil did point out that one problem he faced was persuading tree service companies to look at the trees they were about to fell, not as trunks and limbs to be cut for easy hauling and disposal, but as potential saw logs. Cutting tree trunks for disposal is different from cutting them into usable saw logs. Indeed, trunks cut into logs for removal are not usable as saw logs. Lynn Erickson, a former log buyer for Minnesota Hardwoods, provided simple instructions on how the tree services should cut tree trunks into saw logs. Informal discussions and the occasional presentation were all that was really required. Initially, Lynn estimates, about 10 percent of the logs were usable. However, once the companies learned how to cut, Lynn said that 90 percent were usable.[19]

Lynn's instructions were easy to follow and there was an economic incentive to do so because the companies were paid for the logs they produced and they did not have to pay out as much in disposal fees. Cutting saw logs earned the companies checks from Minnesota Hardwoods and saved them from writing larger checks for disposal.

Another problem is the perception that urban trees are filled with metal and concrete which, when undetected, can seriously damage or even destroy circular sawmill blades. This means repairing or replacing the blade, downtime for the mill, and possible injury to the sawyers. From his own experience, Phil feels that the perception of the problem is somewhat greater than the actual problem.

Lynn feels that metal from trees is a problem, as is concrete and the occasional glass insulator from power lines. From experience, Lynn learned what to avoid. He will not buy double-hearted trees—trees with two or more major stems or trunks—because they are very likely to have large eyebolts embedded in them where the stems had been cabled together as a way of strengthening the tree. Those close to homes were also likely to contain metal—nails used in tree houses and to hold clotheslines. He avoids hollow trees as well because they often contain concrete. Trees felled to make way for development are, in Lynn's view, a good source of logs that are least likely to contain metal.

My own experience over the past six years leads me to a conclusion somewhere between Phil's and Lynn's. I have encountered concrete only once but metal, mainly nails and spikes, several times. Portable band saw mill blades will be dulled by metal, occasionally irreparably damaged (newer bimetal blades are more durable). However, the blades are not expensive and there is relatively little danger to the operator, even if the blade breaks. I have more to say about this issue in the next chapter.

As Phil and Lynn both pointed out, and the abovementioned Task Force also recognized, the prerequisite for using urban trees as a resource requires those involved to stop looking at them as trash or as a serious hazard to their sawmill equipment.

Firewood Instead of Lumber

Another Minnesota city, St. Cloud, started its own program in the 1980s when its elms began dying in large numbers. The City annually cut about 20,000 board feet of lumber from its trees. The lumber was used as posts, benches, parking lot stops, and as sideboards for hockey rinks. By 1988, more lumber was produced than used, so the City switched to making firewood for park buildings and for public sale. Large quantities were also sold to businesses for industrial heating.[20]

According to Dennis Ludivig, urban forester for St. Cloud, the City no longer produces lumber for its own use.[21] It now buys what it needs. Instead, the logs are either sold or are used for firewood. St. Cloud is north of the Twin Cities where the climate is even colder. Dennis said that given the limited variety of trees that grow in the area, mainly elm, oak, basswood and ash, the City makes better use of the wood, more than 200 cords annually, to heat 14 public buildings. What is not used for fuel is mulched and either used in the parks or is given away to the public. Cutting for lumber was also difficult because many of the trees did contain substantial amounts of metal. This is an urban area and a set of circumstances where the public use of trees for lumber does not seem worthwhile.

Urban Forestry in Califonia

AB 939

In 1989, the same year Minnesota passed its Waste Management Act, the California legislature passed Assembly Bill (AB) 939, the California Integrated Waste Management Act. Using 1989 as the base year, AB 939 called for a 25 percent reduction of solid waste by 1995 and a 50 percent reduction by 2000.[22]

Like the Minnesota law, the means to meet these goals were organized hierarchically, starting with source reduction at the top. This requires local governments to find ways of reducing the flow of solid wastes at their sources. Recycling, composting, and reuse are next—as methods of reducing the output that remains. Landfilling was the least preferred way, again the last resort. In 1995, the state exceeded the 25 percent goal by diverting 28 percent of the nearly 50 million tons of solid waste generated that year. In 2000, the state fell short of its 50 percent goal, managing to divert 42 percent of the 66 million tons of solid waste it generated.[23]

The Economic Value of Harvesting Urban Trees

With AB 939 in mind early in the 1990s, Eric Oldar of the California Department of Forestry and Fire Protection (CDF) began promoting two related ideas about trees in California. First, valuable lumber can be cut from trees that would otherwise go to waste or be given away. Eric points out that neither the public nor public officials object to a street tree being cut into firewood lengths and stacked at the curb for anyone who wants it.[24] Yet, objections would be voiced if, for example, the local police department parked a used police cruiser at the curb with the keys in it and a sign taped to the windshield stating that the vehicle was available for anyone who wants it—the point being that urban trees should be treated as having salvage or resale value, just as other tangible assets do that belong to public agencies, or to private property owners.

As I pointed out above, urban logging is the opposite of commercial logging in that we do not get to pick what we want—we must be prepared to want what has been already picked, for either public safety reasons or by natural forces such as storms and disease.

Eric's second idea is that we can come closer, in urban areas, to picking what we want by choosing to plant trees that will have market value as lumber when they are felled. His point is that if we are going to invest public funds over several decades caring for the urban trees we plant, then why not select trees that will provide a return on that investment? He is not advocating the treatment of urban forests as tree farms where large quantities of particular species would be planted solely for their market value—specific growing conditions, environmental considerations, aesthetic appeal, and public preferences would still influence selection. However, there are long-term implications of adding economic value to the list of reasons that govern the selection of urban trees, and these implications apply to both public and private properties.

In the first place, while tree maintenance for safety and convenience would still be priorities, trees would also have to be trimmed in a way that enhances their eventual economic value as saw logs. Basically, this means cutting and trimming to promote the growth of tall, straight, single-stem trees. Trees grown in this fashion are much more likely to yield valuable saw logs.

The selection of species is also influenced by economic value. In the eastern U.S., this means tilting the selection toward walnut, oak, cherry, hard maple, poplar and ash—all hardwoods that have established and substantial market value. Likewise, in the western states, hardwood species like ash, elm, sycamore and oaks would also be favored.

We also should not ignore species that are at the edge of, or out of the commercial mainstream: pecan and osage orange are my personal favorites. Pecan resembles cherry in color and grain and it is a pleasure to work with, and the fruit is edible. Osage produces a hard gold, orange, and yellow lumber that has a striking grain pattern. I would not plant osage near streets or sidewalks, however, because of the inedible softball-sized fruit it drops.

To promote good form and tree health, selection standards for all species should favor nursery stock, either bare-root or containerized, which have a dominant single stem.

Timing of plantings is also affected by the economic factor. A biologically sustainable forest, whether urban or rural, must consist of trees of different ages so that, as the older trees are removed, there are slightly younger ones just behind them in adequate numbers for subsequent removal.

Forests are sustainable as long as the quantity and variety of the forest inventory remains stable from removal to removal. In urban settings, especially in housing and commercial developments, a forest consisting of a variety of trees of different ages is often completely cleared. New trees tend to be the same species and, especially, the same youthful age. What eventually results is more like a tree farm, as opposed to a forest, in that the communities now have a crop of the same trees of the same age. If they are removed due to conflicts with infrastructure like sidewalks, or attacked by pests or disease, suddenly residents find themselves living in a clear-cut neighborhood.

I know from personal experience how this works. My wife, Pat, and I lived in a house on a small urban lot in Cincinnati with six soft maple trees, an ash, a locust, and a magnolia tree. Five of the maples and the locust had been planted at the same time in the early 1900s when the house was built. In one five-year period, as we could afford to do so, five maples and the locust had to be taken down. At about the same time, they reached the end of their natural lives and became hazards to our house and to us. Where we once had nine trees, we were left with two trees, a pile of tree removal bills, and no lumber.

Seeing Urban Trees as an Urban Forest

The final issue raised by Eric's approach to sustainable urban forestry is one that is easy to see from the air but would be a challenge to put into practice on the ground. In a summertime flight, I have seen my own and surrounding properties from a helicopter. From this bird's eye view, I can see the forest collectively and have a greater sense of how all the trees and other vegetation, mine included, make up the urban forest in my city.[25]

On the ground, in our yard, I saw just our own and our neighbor's trees. That sense of my trees as part of the larger urban forest is hard for me to see, literally and figuratively, when I'm standing in my backyard thinking about my property and, like most Americans, my right to plant the trees I want and remove them only when I must.

Eric argues that on the ground we should try to think of the trees planted at our homes and businesses as part of the whole urban forest. From the bird's eye view, practicing biologically and economically sustainable urban forestry would require each of us to plant and remove with the total forest in mind, as opposed to the narrower perspective of individual owners.

For each of us, this means realizing that the trees in our yards and on our streets are not immortal. We must seek a better balance in our planting and removals, both as individuals and communities, to ensure that when the time arrives to harvest a tree, a replacement will be right there ready to take its place. We must come to a point where we understand that the best time to remove trees is not when they are stone cold dead, fatally diseased or insect infested, but when they have reached their economic maturity. Harvesting trees in this fashion will ensure that they financially support their replacements and, on a larger scale, support sustainable community forests.

The long-term goal of such a policy would be to create a biologically sustainable urban forest that would be economically self-sustaining as well. The revenue from the sale of lumber, logs, and green waste used for firewood and mulch, plus funds not spent on landfill expenses could, when added together, support the purchase and planting of new trees and the maintenance of the existing urban forests.

Whether enough urban Americans, and

their communities, could ever adopt such a broad point of view depends at least in part on the prevailing sense of community. This sense would have to be based on the feeling that what everyone does together adds up to more for each individual than what each would have without the collective effort. Since roughly four-fifths of urban property is privately owned, a good number of owners would have to participate to make this a workable long-term plan for the entire community forest. Eric argues that if most sign on to such a plan, then the sustainable community forest could generate enough revenue from the sale of logs and other tree material and from reduced disposal costs to purchase, plant and subsidize tree maintenance on private property.

This approach could be characterized as active urban forestry in the sense that there would be a comprehensive policy for the entire urban forest that recognizes the economics of tree planting, maintenance, and harvesting on both public and private properties. The alternative more passive approach we see in, for example, New Jersey, Minnesota, and my own effort in Ohio, is to take what there is and do the best we can with it.

Whether an active approach could be translated into public policy is open to question. From my own experience, I suspect initial resistance would arise from the need to think of one's own trees as part of the larger forest and as a plant with market value as opposed to a living thing that has become part of our lives. Unlike we do with other vegetation, we tend to think of trees—especially the ones we have lived with for a while—more as pets than as plants. Still, I would like to think that one state, possibly California, and some of its communities, could try. We could be driven to the reclamation of urban trees by the economics of waste disposal restrictions, expenses and tree availability, as we are now, or we could drive the economics ourselves by adapting a more active policy. An active policy would still include waste disposal but we would not be as dependent on it as we are at present.

The CDF Sawmill and Kiln Loan Program

In the meantime, California moves forward a few steps at a time. The first step was taken in 1996 when the CDF purchased its first Wood-Mizer® portable band saw mill and a kiln. Since then four more mills have been acquired, along with four new mobile kilns. Inside California, CDF makes this equipment available to local governments, non-profit organizations, and private firms willing to explore a new entrepreneurial venture. Cindy McCall, manager of Parks and Urban Forestry in Lompoc, was the first to take advantage of the offer (described in Chapter Four) as was West Coast Arborists, Inc., a tree service company in Southern California. The Riverside Corona Resource Conservation District has also used the equipment. At present, a mill and kiln are on loan to Palomar Community College's Cabinet and Furniture Technology Program (see some examples of student work below). Another agreement is being processed for Protect All Life (P.A.L.) for use of equipment to reclaim hardwoods in the San Francisco Bay Area.

FIGURE 7
Zuni Bear clock made by Dennis Strand, a student at Palomar Community College's Cabinet and Furniture Technology Program, from Red Iron Bark eucalyptus and sycamore.
Photo by Archie Breeden

FIGURE 8
Nightstand in Black Acacia also made by Dennis Strand.
Photo by Archie Breeden

FIGURE 9
Steel string guitar made by Stuart Austin, also a Palomar student, from Red Gum eucalyptus, Black acacia, and Sitka spruce.
Photo by Archie Breeden

Saw It and They Will Buy?

"Who will buy urban lumber?" is a central economic question. One option is to move logs and lumber into the existing commercial log and lumber market. Both Minnesota and New Jersey tend to favor this route. Buyers and sellers in this market routinely deal in large quantities (ten thousand board feet, or about a tractor-trailer load) that must be graded to industry standards. Grading is essential to buyers of large quantities, whether they are dealers or end-users such as large furniture or flooring manufacturers. Since none of them could inspect every board they buy, they must rely on standards that yield a homogeneous commodity. For American hardwoods, prevailing market prices for the different grades are published on a regular basis in the *Hardwood Market Report* and the *Weekly Hardwood Review.*

Park officials, tree services, utility companies and developers together in large urban areas might be able to generate enough logs and lumber to sell on a regular basis. Small communities, firms and community organizations may find the demands of this market to be more than they can handle.

For sellers of urban wood, there are two disadvantages of selling in the commercial market. You must be prepared to offer large quantities on a regular basis and you must be prepared to accept prices based on industry grading standards. Those standards are based on the presence of defects such as knots that reduce the amount of clear wood that can be cut from a board. Defects that reduce the grade of lumber also reduce its price.

However, a defect in the commercial market might not be a defect to custom cabinetmakers and to woodworkers. Selling to this more specialized segment of the wood market is a second option. Wood that is unacceptable in the commercial market, spalted maple for example, or highly figured limb wood, would still be highly prized for use in custom cabinetry and furniture. This also applies to wood not ordinarily found in the commercial market, such as carob in California or osage orange and pecan in Ohio. Because of cracking, cupping, and warping during drying, boards cut from limbs would be judged defective or suitable only for pallets.

From my own experience, I know this can be so. My son and I have made 20 different pieces, from bookcases to rocking horses, from the massive limbs of a bur oak estimated to be about 500 years old when it fell in 1996 (more on this in Chapter Four). The sawyer who kiln-dried the wood told me that, at best, the lumber was suitable for pallets. He was thinking in terms of its commercial value, not about the oak's unusual grain or the personal value of the wood to its owners. For example, many of the boards chosen by George Nakashima for the unique pieces of furniture he created would not pass the grader's eye. A split down the middle of a black walnut board is a commercial defect but to Nakashima it was an opportunity to use a butterfly key as a way of turning the split into an important visual feature of the piece[26] (see Figures 13 and 14). Perhaps an additional grading standard is required to recognize at least some characteristics, deemed flaws by the current commercial standard, as valuable attributes.

According to Eric, efforts in California presently favor the specialty market, looking for woodworkers, turners, carvers, custom furniture and cabinetmakers in the market for wood with unusual grain pat-

FIGURE 10 Timber frame building, constructed by students and faculty at Palomar Community College, that houses the Wood-Mizer® mill on loan from the CDF. **Photo by Lisa Pieropan**

terns—so unusual as to probably fail the commercial lumber grading standards. Attending shows and conferences around the state, he has found, are effective ways to inform potential buyers of the virtues and value of unusual urban wood.

Of course, a third option is to sell to both markets as the opportunities arise. When enough logs of commercial value have accumulated, they can be sold to sawmills. The very best can be sold to veneer mills. If the sawmills are unwilling to accept urban logs, then portable band saw mill operators could cut them instead. The wet lumber could be graded and sold into the commercial market. Or, if a kiln is available, the wood can be dried and then sold at higher prices. Meanwhile, unusual wood that would not pass the grading muster can be sold into the specialty market.

Summary

As experience in New Jersey, Minnesota, and California demonstrates, recent efforts to reclaim urban trees for their lumber are driven by broader governmental actions to reduce the solid waste generated by the

FIGURE 11
One of the CDF portable kilns.

FIGURE 12
A kiln being loaded for drying.

nation's cities. Additional impetus comes from the growing awareness that we need to make more efficient use of our resources by reusing, in the widest meaning of this word, what we produce and consume.

The essential first step, already underway, requires us to look at urban trees as a resource, not as garbage. This means that those who deal directly with the planting, care, and removal of urban trees—tree service companies, parks and recreation officials, street maintenance officials, arborists, and landscape designers—must see the trees as a resource even before they are felled. Cutting in short lengths for ease of removal at most creates firewood, material for mulching, or waste.

Cutting trees into saw logs creates the opportunity to derive more value from them—in some cases, much more value than cutting them into firewood or grinding them into mulch. New business opportunities exist for tree service companies willing to cut and sell logs and lumber. Not only is this a potential source of additional revenue, it is a major way to reduce disposal costs. For these companies, such costs are substantial and rising.

Local parks and street maintenance agencies can reduce their own operating expenses, and create new sources of revenue for city parks and streets the same way. Lumber from public trees can be used to make public furniture and for steps, rail-

FIGURE 13
Here an East Indian rosewood butterfly key is used on a cracked knot in a cabinet made of claro walnut.
Photo by David Conover of George Nakashima Woodworker, S.A.

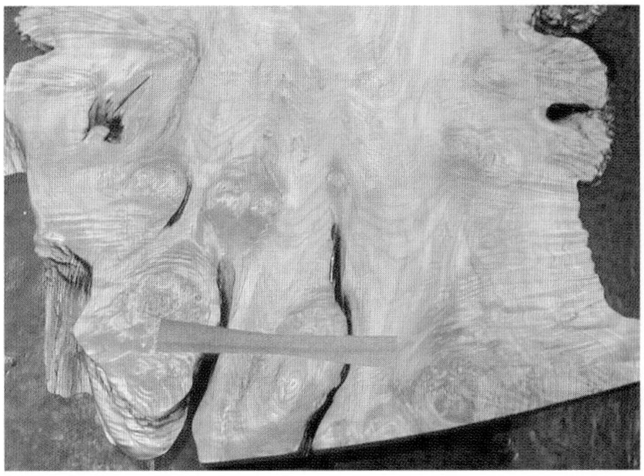

FIGURE 14
A butterfly key made from East Indian laurel spans two bark inclusions in a tabletop made from an Oregon maple burl.
Photo by David Conover of George Nakashima Woodworker, S.A.

ings, parking lot blocks, and even construction. Public funds are thereby saved, and an example is set for others to follow.

Tree service companies, public agencies, and community organizations must decide which of two market options works best for them: the commercial or the specialty market. A third option is to sell into both as the opportunities arise.

The immediate consequence of cutting urban trees into logs and lumber is that we can reduce both public and private costs of disposal while acquiring additional value from the trees beyond their use as firewood and mulch. The larger issue raised by this effort goes to the heart of how we think of our urban forests. On the one hand, we can continue to think of them primarily as valuable urban amenities whose costs of maintenance, removal, and disposal we must bear. Or, by introducing economic value into a long-term strategy on what and where to plant and when to cut, we can move toward an urban forest that is more likely to be both sustainable and economically self-sustaining.

Regardless of the specific approach we take, the nation's urban forest must be able to survive both the biological conditions imposed by nature and the economic forces we impose.

FIGURE 15
"Defective" edge becomes attractive feature in table made from reclaimed California hardwood.

2 FROM TREES TO LOGS

Introduction

The first step in our effort to make better use of urban trees is to recognize them as a potential source of valuable lumber, not just as mulch, firewood, and green waste. The second step is to turn fallen trees into logs. To do this, we need to know something about the trees themselves and what is involved in felling and cutting them into logs.

We are helped at both steps if we also know how many urban trees there are: that is, what is the inventory? More specifically, we need to know the age, species, and condition of the inventory. We also need to know how many are removed every year and why, and how many are added and what kind. This information is essential to estimating the number, quality, and kind of logs we can expect to get from the urban tree inventory both nationwide and for particular urban areas.

We can cut logs from trees without knowing anything about trees themselves. However, knowing something about how they are constructed, and especially how they grow, will enable us to obtain more usable and higher quality lumber, avoid damage to saws, and, most importantly, injury to ourselves.

Safety is the most important issue to keep in mind when trees are being brought down and cut into logs. Knowing what is involved, the techniques and the dangers, of felling, limbing, and cutting trunks and limbs into saw log lengths (known as bucking) will minimize the risk of injury. As I point out in the Introduction, cutting trees is risky business. Not only are felling, limbing, and bucking dangerous chores in their own right, so are the chain saws we use to do this cutting. What might seem to be a minor oversight or a momentary lapse of attention can quickly lead to a major injury, or worse.

In the previous chapter, I discussed the reasons to utilize urban trees as a source of lumber and presented three important cases where this is being done.

This chapter describes the process of getting from trees to logs. Taking this step means examining what we know about the urban tree inventory and changes in that inventory. We need to have a basic understanding of the structure and growth of trees. We need to distinguish between trees that are worth salvaging and those that are not. Finally, but very importantly, we must know and follow safe cutting methods.

The next chapter describes how logs are cut into lumber and the chapter after that describes how the lumber is then dried.

The Population of Urban Trees

What Do We Know?

As required by the U.S. Constitution, every ten years the Population Census provides us with a wealth of new information on the social and economic characteristics of the American population. From the Census data for American urban areas, we learn the size of urban populations and how those populations are distributed by such characteristics as age, education, income, ethnicity, and race. Comparisons of decennial census results, along with sample surveys conducted during the ten-year intervals, tell us how populations are changing in size and composition.

There is equivalent information on the status and condition of the nation's nonurban forests collected by the USDA Forest Service's Forest Inventory and Analysis (FIA) program.[1] Available by state, the inventory is scheduled to be updated every five years. This information allows us to monitor changes in the size and composition of the nation's forests. Though we certainly need these numbers, at present we know more about the trees in rural forests than we do about those on our streets and in our backyards and parks. However, the Forest Service is currently pilot-testing methods for a national program of urban forest inventory and health-monitoring.

Efforts to reclaim urban trees for lumber would be greatly helped if we had precise urban tree counts as well as estimates of the number of plantings and removals. Knowing the number of trees removed, especially by age, species, and the reasons for removal, would enable us to more precisely determine the level of effort and organization needed to reclaim these trees for lumber, and the economics of doing so. Plantings, together with removal numbers, would allow us to make projections about the future size and composition of individual urban forests, and the outlook for future tree reclamation efforts. Other than in a dozen urban areas, this information is not available yet.[2]

Nationwide Urban Forest Survey

The USDA Forest Service recognizes the need for a nationwide urban forest inventory, equivalent to the ongoing non-urban inventory, to support urban forest management. Using satellite imagery (with a one-kilometer resolution) to measure forest density, the Forest Service has for the first time estimated the tree coverage for urban America.[3] Both the coverage and estimated tree count at the state level for metropolitan and urban areas are in given in Table 1 below.[4]

As explained in the Introduction, the federal government uses two different but somewhat overlapping geographic definitions of the broad term "urban". For the 1990 census, the Bureau of the Census defines urbanized areas as places with a minimum population of 50,000 and a density of 1,000 people per square mile. Urban places have a minimum population of 2,500 and must be outside urbanized areas. Places with fewer than 2,500 residents outside urbanized areas are considered rural. For example, Atlanta (population 416,474) is an urbanized area while Valdosta (population 43,724) is an urban place. Plains, Georgia (population 637) is rural (population statistics are for the year 2000).

The Office of Management and Budget (OMB) defines major urban populations and centers by counties, not places (except in the six New England states). By the OMB definition, metropolitan areas have one or more large core populations that are socially and economically linked to adjacent counties. For example, Cincinnati residents are the core population for the 12 county Cincinnati metropolitan area whose current population is about 2 million. The 20-county Atlanta metropolitan area had a population of about 4 million. The nation's largest metropolitan areas are comprised of several large population centers and adjacent counties all linked together. With a population of over 21 million, the largest is the New York metro area that covers 36 counties in New York, Connecticut, New Jersey, and Pennsylvania.

The OMB definition covers about seven times more territory than the Census definition. In Table 1, the total amount of land covered by metropolitan areas ("MA" under the "Areas" column) is 764,164 square miles (1,979,700 square kilometers). The total land covered by urban areas ("UA" under the "Areas" column) is 168,466 square miles (281,000 square kilometers). Of course, these two different definitions yield very different estimates of the metropolitan/urban tree populations. There are an estimated 74.5 billion trees in the nation's metropolitan areas ("MA" under "Tree Population"). There are just under 4 billion in urban areas ("UA" under "Tree Population").[5]

While metropolitan areas cover about seven times more territory than urban areas, metro areas have almost twenty times the estimated number of trees. My own city and county residence illustrates why this is so. I live in Milford, Ohio (2000 population of 6284), which is in the western part of Clermont County and close to Cincinnati, Ohio (in adjacent Hamilton County). By the Census definition, only the trees in Milford and a few other urban places in Clermont County would be counted.

However, Clermont is one of the dozen counties that comprise the Cincinnati metropolitan area. By the metro definition, all the trees in Clermont County are counted as urban trees even though the eastern side

TABLE 1

Estimated Number of Urban Trees, Tree Cover, and Urban Area by State

State	Area		Area of State[1]		Tree Cover[2]		State Tree Cover[3]		Tree Population[4]		Trees/Capita[5]	
	MA	UA	MA	UA	MA	UA	MA	UA	MA	UA	MA	UA
	Square Kilometers		---------- Per Cent ----------						--Thousand Trees --			
Alabama	44,950	8,487	33.1	6.3	57.5	48.2	29.7	4.7	3,767,917	205,847	1,391	69
Arizona	159,133	9,218	53.9	3.1	12.9	11.4	45.5	2.4	633,609	53,950	198	9
Arkansas	23,023	3,435	16.7	2.5	30.6	25.0	12.3	1.5	910,335	43,412	821	32
California	253,807	27,348	59.9	6.4	20.6	10.9	39.2	2.2	2,134,318	148,612	74	5
Colorado	49,353	4,345	18.3	1.6	20.5	13.0	15.1	0.8	738,130	28,149	266	7
Connecticut	10,561	4,085	73.6	28.5	39.1	21.8	65.2	14.0	431,348	44,800	143	14
Delaware	3,351	566	52.0	8.8	50.9	46.3	58.2	9.0	212,596	13,257	384	27
Florida	97,442	18,407	57.2	10.8	31.1	18.4	49.5	5.5	3,459,653	169,587	288	13
Georgia	34,339	8,338	22.3	5.4	68.1	55.3	24.0	4.7	3,442,105	232,906	791	49
Idaho	7,282	966	3.4	0.4	23.7	25.6	1.9	0.3	133,053	12,494	368	18
Illinois	48,489	9,165	32.3	6.1	38.6	33.7	33.0	5.5	1,951,744	155,544	204	14
Indiana	36,324	5,000	38.5	5.3	36.5	31.2	35.7	4.2	1,480,599	78,498	374	21
Iowa	17,092	3,148	11.7	2.2	36.7	33.1	11.5	1.9	626,272	52,474	522	29
Kansas	14,819	2,575	7.0	1.2	21.5	20.5	17.4	2.9	333,535	26,677	250	17
Kentucky	17,216	3,374	16.4	3.2	42.7	33.4	12.5	1.9	839,464	56,681	472	23
Louisiana	53,818	5,374	40.1	4.0	38.2	25.3	36.0	2.4	2,507,970	68,510	794	19
Maine	13,651	2,887	14.9	3.1	64.7	47.7	14.0	2.2	1,942,124	68,550	3,924	110
Maryland	18,340	4,525	57.1	14.1	46.5	40.1	53.2	11.1	850,811	89,434	192	21
Massachusetts	23,403	6,893	85.6	25.2	45.1	25.3	88.4	14.4	1,258,989	86,829	212	17
Michigan	54,971	7,494	21.9	3.0	37.4	29.7	15.0	1.6	2,486,540	110,858	323	17
Minnesota	46,555	6,775	20.7	3.0	50.0	37.4	20.2	2.2	2,795,672	127,767	929	33
Mississippi	16,747	3,365	13.3	2.7	54.8	38.6	12.9	1.8	1,287,176	65,520	1,472	48
MIssouri	31,838	5,655	17.6	3.1	36.1	30.6	15.3	2.3	1,323,812	87,148	379	21
Montana	13,885	4,365	3.6	1.1	16.9	49.4	2.4	2.2	140,484	108,550	735	251
Nebraska	6,892	1,061	3.4	0.5	29.8	21.1	8.0	0.9	164,718	11,243	209	10
Nevada	84,958	3,195	29.7	1.1	12.1	9.9	26.5	0.8	129,004	15,834	127	9
New Hampshire	5,362	1,678	22.1	6.9	58.8	49.1	17.7	4.6	453,046	41,455	660	60
New Jersey	22,590	6,916	100.0	30.6	56.6	41.4	100.0	22.3	1,596,644	143,869	207	20
New Mexico	30,528	2,316	9.7	0.7	11.9	4.8	9.0	0.3	264,298	5,682	314	4
New York	72,841	10,127	51.6	7.2	44.7	26.3	43.9	3.5	4,597,839	132,466	278	8
North Carolina	48,529	6,419	34.8	4.6	52.5	42.9	31.4	3.4	4,356,839	138,606	996	36
North Dakota	17,667	457	9.6	0.2	14.3	7.8	11.8	0.2	148,126	1,774	575	5
Ohio	55,178	9,923	47.5	8.5	44.7	38.3	45.6	7.0	3,249,536	191,113	368	22
Oregon	36,108	2,280	14.4	0.9	75.1	30.4	22.6	0.6	2,138,017	34,583	1,077	17
Pennsylvania	58,315	8,363	48.9	7.0	48.7	34.4	43.5	4.2	3,732,947	139,020	370	16
Rhode Island	3,189	926	79.7	23.2	38.3	8.9	88.9	6.0	129,074	4,155	141	5
South Carolina	33,670	4,380	40.6	5.3	54.4	39.8	38.5	3.6	2,986,395	86,696	1,233	44
South Dakota	10,818	617	5.4	0.3	25.1	19.2	10.6	0.5	17,682	6,007	80	15
Tennesee	33,420	7,382	30.6	6.8	53.8	43.9	28.1	5.1	2,404,114	163,783	726	49
Texas	143,184	26,573	20.6	3.8	15.8	10.5	29.4	3.6	2,755,780	140,709	195	8
Utah	21,628	2,577	9.8	1.2	19.5	14.0	11.4	1.0	209,934	18,330	157	9
Vermont	3,903	416	15.7	1.7	48.9	36.0	10.5	0.8	242,078	7,558	1,367	42
Virgina	40,759	8,869	36.8	8.0	53.3	35.3	34.4	4.9	3,647,767	156,545	764	27
Washington	52,953	5,679	28.7	3.1	50.9	33.6	28.6	2.0	2,089,982	93,272	518	23
West Virgina	10,108	1,086	16.1	1.7	65.6	42.2	13.4	0.9	890,930	22,871	1,191	33
Wisconsin	43,747	4,565	25.8	2.7	39.0	25.8	21.8	1.5	1,886,180	59,344	566	18
Wyoming	20,885	797	8.2	0.3	4.1	3.6	2.0	0.1	43,464	1,392	323	3
United States	1,979,700	281,000	24.5	3.5	33.4	27.1	24.5	2.8	74,425,644	3,820,491	377	17

of the county, furthest from Cincinnati, is more rural than urban and is less under Cincinnati's social and economic influence than the western side. And, there are more trees and more wooded areas and small forests in the eastern side.

In general, while by definition all metro trees are urban trees, when it comes to harvesting, they are not all the same. As discussed below in the section on felling trees in urban areas, the significance of this point has to do with the number of trees involved, the way they are harvested, and by whom.

While we have estimates of the urban tree population, determining what species they are, and their size and condition is still beyond the current capability of satellite imagery (and may remain out of reach for another decade). For our purposes, the satellites can tell us the size of the urban forest but little of its content.

However, specific and accurate numbers on content can be obtained from on-ground surveys. Baseline counts have already been conducted in about a dozen American cities (plus two in Canada and one in Chile). Since counts have just begun, there is as yet no data on change in the size or composition of the forests in these cities, except for Syracuse, New York, and Baltimore, Maryland where a baseline count was done in 1999 and a second count in 2001.[6] We need the numbers from at least two years in order to measure and explain changes in the inventory.

In general, the more numbers we have over time, the more we will know about how urban forests are changing and in what direction. This information can be used to estimate the more immediate economic value of urban trees as amenities, as an important part of the urban infrastructure, as a source of lumber, and to understand the value of the forests' capacity to cleanse the air of pollutants we continue to generate, especially in metropolitan areas. On a longer-term basis, we can estimate the value of their ability to sequester (that is, pull and hold) carbon dioxide from the atmosphere, thus helping to moderate global warming.

Neither the currently available information, nor the information that will be available in the near future, is being collected for the express purpose of measuring the number of logs that urban trees might yield. This means that log volume numbers will have to be based on estimates of how many trees are removed each year and what proportion of those could yield logs. The quantity of lumber will

Table 1 (facing page)

Estimated Number of Urban Trees, Tree Cover, and Urban Area by State

1 Percentage of the state's total area that is occupied by metropolitan areas (MA) and urban areas (UA).

2 Percentage of metropolitan areas (MA) or urban areas (UA) covered by tree canopies.

3 Percentage of total state tree cover within metropolitan areas (MA) and urban areas (UA).

4 Includes the District of Columbia, but not Alaska or Hawaii.

5 Includes 482 square kilometers of urban area (UA) that crossed state borders and could not be assigned to individual states.

Source: Dwyer, John F., David J. Nowak, Mary Heather Noble, and Susan M. Sisinni, August, 2000. Connecting People With Ecosystems in the 21st Century: An Assessment of Our Nation's Urban Forests, U. S. Department of Agriculture, Forest Service, Northeastern Research Station, Syracuse, NY., p. 24 - 25. Though gathered at somewhat different times in the decade of the 1990s, these data are estimates for 1991.

have to be estimated from the log quantity estimate. In short, log estimates are one step removed from the available data and lumber estimates are two steps removed. To calculate the future flow of logs and lumber, we need to know the planting rate and composition of what is being planted. Together, removal and planting numbers will then allow us to estimate the quantity and type of logs and lumber to expect in the future.

Local and State Sources of Information

In the meantime, without detailed urban tree inventories, we will have to rely on urban foresters, urban park personnel, state departments of natural resources, and tree-service arborists for information on the species of trees in each of our cities. Park and natural resource employees will be most informed about public trees while tree services will know more about what is growing on private property. Though none will have a comprehensive count they will, by training and local experience, know the indigenous species and should have a sense of the proportions of each in the local urban forest. They will also know of trees not native to their areas that are, nevertheless, valuable.

The Structure of Trees

Understanding how nature builds trees helps us understand why some urban trees are worth the effort to reclaim and others are not. Knowing the difference allows us to focus our limited resources on those that are worth saving for lumber. Understanding the structure of trees also helps us later on, when we prepare the wood for drying and then for its final end uses.

As I indicated in the Introduction, trees are woody perennial plants with either one or just a few main stems. At full growth, they are generally larger than other plants such as shrubs (which are also woody perennials, but with many stems).

Trees are both the largest and the oldest living things on earth. By volume, the largest known tree is the General Sherman Giant Sequoia in California. At about 275 feet in height and 36 feet in diameter, it contains an estimated 52,500 cubic feet of wood (or about 427,000 board feet). It weighs an estimated 6,600 tons and is probably between 1,800 and 2,700 years old.[7]

The oldest living tree found so far is a bristlecone pine, named Methuselah, growing somewhere in California's White Mountains. It is less than 30 feet in height but is estimated to be about 4,800 years old.[8]

Softwoods and Hardwoods

In non-botanical terms, trees are broadly divided into softwoods (conifers) and hardwoods (broad-leaved trees). Though these two words imply that hardwoods are harder than softwoods, there are enough exceptions to make this at best a loosely applicable generalization. A familiar example is southern pine (which, commercially, consists of longleaf, short-leaf, loblolly, and slash pine) used in pressure-treated form to construct outdoor decks and other weather-resistant structures. It is harder than poplar, a common hardwood used to make indoor furniture and trim molding.

Softwoods are usually evergreen and retain their needle-like leaves for more than one year (bald cypress and larch are among the few exceptions). Evergreen examples are pine, fir, redwood, cedar, and cypress trees. Hardwoods in the temperate areas of the U.S. are usually deciduous: they shed their broad leaves each fall and re-grow them the following spring (American holly is an exception). Walnut, oak, ash and maple are common examples in the eastern U.S.

In botanical terms, both hardwoods and softwoods are classified as spermatophytes, a division of the plant kingdom that encompasses all seed plants. Hardwoods are angiosperms ("angio" from the Greek root *angeion* meaning a case or capsule), the immature seeds of which are enclosed within the ovary of the flower. The mature seeds are contained in the fruit of the tree. Acorns, walnuts, hickory nuts, peaches and maple keys ("whirligigs") are familiar examples. Softwoods are gymnosperms ("gymno" is from the Greek root *gymnos* meaning "naked seed"), the seeds of which are not enclosed and are most often found in cones (hence, "conifer").

Tree Growth

Leaves grow from twigs and collectively form the crown of the tree. From a side view, the outline of the crown tends to be umbrella-like, with lower limbs growing laterally and higher limbs growing vertically. This is, however, a theme with many variations. The exact shape is determined by several influences, including the exact place where the tree is growing. Those growing close to one another in forests tend to be taller and have somewhat smaller crowns than those growing alone in open spaces or at some distance from nearby trees. Softwood trees have a main trunk, with branches growing out laterally. From the side, the outline of the mature softwood tends to be triangular, an elongated version of the usual Christmas tree shape.

Four hundred feet seems to be the maximum height any tree can grow even under the most favorable conditions. A tree cannot exceed this height because its cells cannot grow under the water pressure that would be required to move moisture between roots and leaves. In addition, structural problems arise from the sheer weight of the trunk as it thickens at the base to hold the proportionally greater weight that comes with heights above four hundred feet.[9]

The roots of a tree grow down and out horizontally into the ground, thereby structurally anchoring the tree to the earth. Root tips pull water from the ground and are the front end of a pressure system that helps push the water in a continuous molecular stream through the vessels in the stem, limbs, and twigs all the way to the leaves (some water also enters through the woody part of the roots as well). In the leaves, photosynthesis, a process driven by the energy of sunlight absorbed by chlorophyll, creates carbohydrates for the tree from carbon dioxide and water. Oxygen is produced as a by-product of this process. Almost all of the water drawn to the leaves evaporates from pores that open to the surfaces of the leaves. This evaporation, or transpiration, cools the leaves and the tree.

If trees are close enough (and there are

enough of them), transpiration, wind, and shade combine to cool our homes and offices and thereby reduce the electrical energy needed to run air-conditioners.[10, 11] From the perspective of our own self-interest, healthy trees are oxygen-producing air-conditioners that cost very little to run, and do little or no environmental damage. To the contrary, they tend to repair the damage we do to our own air by swapping oxygen for the carbon dioxide we produce (the slightly odd term for this is the "sequestration of carbon").[12]

Structure at the Cellular Level

As shown below in Figure 2.1, concentric layers of cells, all but one visible to the eye, together make up the trunk of a tree. Starting on the outside of the outside ring is the outer bark, the only part of the trunk we usually see. It consists of dead cells on the outside, analogous to the epidermal cells of our skin. The phloem, the layer of inner bark, consists of live cells that serve as the conduit for sap that flows from leaves to roots. Some of the sap moving downward is diverted through horizontal ray cells that run from the phloem to the very thin cambium layer of cells, the next layer in and the one that can be seen only with magnification. The rays continue inward from the cambium layer to the sapwood and heartwood of the tree; the latter two comprising the xylem, or what we commonly think of as wood. Hormones (such as auxin produced by leaf buds) and sugars in the downward moving sap are essential to the maintenance and growth of new cambium, bark, and sapwood cells. The cambium layer produces cambium cells as well as more phloem cells in an outward direction and more sapwood cells toward the center of the tree.

As cells are added horizontally, the girth of the tree expands outward, enclosing whatever it encounters. Limb buds and remnants of broken limbs are enclosed, as are barbed wire fences, metal posts, nails and spikes hammered into the bark, and any other stationary object the tree encounters as it grows ever outward. See Figure 17 for an unusual example. This is an important issue in salvaging urban trees. Because the trunk is expanding out and not up, enclosed objects remain at the height where the tree first encountered them. Knowing this tells us where to look for embedded objects when cutting lumber, usually at eye level (the exception being steel eyebolts used to cable multiple stems or limbs together for support).

After the cambium comes the light-colored sapwood, the outer layer of xylem. This layer varies widely in width: it can be thin in some trees such as black locust (see Figure 18) and thick in others, such as hard maple and southern yellow pine. Even in a given species, sapwood width can vary widely depending on the growth rate of the tree. In general, faster growing trees have wider bands of sapwood cells. They serve as a storage place for nutrients and also as a conduit for mineral-laden sap that moves from the roots to the leaves.

The inner layer of the xylem is the heartwood, very often though not always darker in color than the sapwood. This layer consists of inactive and dead sapwood cells bound together by the glue-like lignin. Heartwood cells are not involved in either the movement of water or food storage but they do provide structural support for the tree. The darker color comes from extractive substances that accumulate in the cells. Among the extractives are tannin, oil, fats, resins, waxes, and colored materials (the word itself refers to the fact that these substances can be removed or extracted from the heartwood by water, alcohol, and other solvents).

Because of the extractives, heartwood dries at a slower rate than sapwood. However, once dry, heartwood is less affected by changes in the ambient humidity. Furniture parts, turned bowls, and other crafted items made from heartwood are less likely to destructively expand and contract with seasonal or weather-related changes in humidity. The width of the heartwood varies inversely with the width of the adjacent sapwood. A fast-growing walnut tree will have more light colored sapwood, and less dark heartwood, than one that grows more slowly.

Heartwood is usually the part of the tree cut for lumber, especially in species such as walnut where there is a very sharp distinction between the dark brown heartwood and the off-white sapwood. Hard maple is one exception where the off-white sapwood is more desired for flooring and cabinetry than the darker heartwood.

Extractives such as tannin in oak provide the heartwood with resistance to decay and to attack by insects and fungi. Tannin in oak lends a desirable vanilla-like flavor and aroma to wines aged in toasted oak barrels, especially those constructed of white oak from the Nevers and Limousin

FIGURE 16
Layers of different cells from bark to heartwood in the trunk of a tree.

areas of France (it must be especially desirable since the fifty-nine-gallon French barrels sell for seven hundred to one thousand dollars each compared with three hundred to five hundred dollars for barrels made from American white oak).

Finally, the pith is at the center of the tree trunk as well as at the center of its branches and twigs. It is the bull's-eye at the center of the growth rings. Because it consists of the first cells to grow in the tree when it was a sapling, the wood immediately around it is referred to as juvenile wood. The rest of the tree stem, branches, and twigs are adult wood. Cracking emanates from the pith outward in a log, often within an hour of felling. Look again at Figure 18 below, this time for the cracks that start in the pith of the locust log. On occasion, I see treated pine posts containing the pith marketed in home centers as the best buys for decks and other projects. As they dry, they will nearly always split and warp. Whether posts or boards, I avoid lumber containing the pith and therefore the juvenile wood from the sapling.

In the temperate climate that prevails throughout most of the U.S., trees go

FIGURE 17
Crucifix partially enclosed by a tree growing near Prague, the Czech Republic.
(Photo by Luciano Duse, photographer.)

through an annual cycle of growth and dormancy. During the growing phase of this cycle, trees add a new ring of wood consisting of two layers. The first layer is called earlywood (or springwood), the second latewood (or summerwood). In trees where the two layers are clearly visible (southern pine, for example) earlywood is lighter in color and softer, while latewood is darker and harder. Counting growth rings, in trees where they are visible, is a way of estimating the age of a tree. Variations in the width of the rings also provide indirect evidence of changes in conditions that affect growth, such as annual weather extremes and long-term climatic shifts.

The wood cell itself is like a long narrow room with small windows, referred to as either pits or perforations, at each end. Arranged in pairs, these openings are positioned directly across from one another in adjacent cells. The cavity formed in the cell by its multi-layered walls is referred to as the lumen. Sap is both held in, and transported through, cell lumens. The sap passes through the paired openings from cell to cell as the tree nourishes itself, or as it dries out after being sawn into lumber.

Over 90 percent of the cells in softwood trees are long, thin longitudinal tracheids packed parallel to one another in straight rows around the tree trunk. The cell ends are tapered and rounded and have multiple pits that match up to those of adjacent cells. Like a pane of glass, a permeable membrane separates the pits of adjacent cells. The tracheid cell has been likened to a soda straw that has been pinched shut at both ends.[13] Like a straw, the cell's length is much greater than its diameter—about a hundred times greater. In many softwoods, such as pine, larch, spruce and Douglas fir, another kind of cell forms canals in the spaces between the tracheids. These cells

FIGURE 18
Cross-section of a locust log. There are two layers of bark: the dead cells on the outside and the living phloem on the inside. Next, the narrow band of lighter colored sapwood surrounds the darker heartwood and the annual growth rings. Cracks caused by drying emanate from the center or pith of the log.
Photo by Sam Sherrill

can defensively secrete resin into the canals when the trunk or a limb is physically penetrated, damaged, or is being invaded by insects.

The resin, distinct from the tree's sap, oozes out, sealing the wound or capturing the insect on its sticky surface. For centuries, pine trees have been tapped for resin, which yields turpentine when distilled. Rosin, a byproduct of distillation, is rubbed on the bows of stringed instruments such as violins to create more resonant tones. Resin itself has also been used as a finish for violins. Ancient insects are sometimes found in amber, which is fossilized resin (the premise of the film, Jurassic Park, was that crystallized blood extracted from mosquitoes trapped in amber was the source of dinosaur DNA). It is used medicinally, aromatically, and even to flavor a Greek wine called Retsina. Despite its many uses, resin can pose a problem when pine lumber is being dried, especially air-dried. This is discussed in the section below on drying lumber.

Perhaps because they evolved more recently than softwoods, hardwoods are composed of a variety of differently shaped cells that are not arranged as neatly as the longitudinal tracheids in softwoods. Among the most notable are vessel element cells, not found in softwoods, that account for twenty to sixty percent of the cells in the sapwood.[14] Linked end-to-end, these cells form tubes or pipelines, known as vessels, in the sapwood. Sap moves through these vessels between roots and leaves. Unlike the pits of the softwood tracheids, the windows of the vessel element cells are open perforations not covered by a membrane. Vessel element cells expand in diameter as they mature, pushing adjacent cells out of an otherwise more linear alignment. This nonlinear arrangement provides the tree's crown with moisture from many of the tree's roots, a redundancy that insures a more reliable water supply for the crown in case roots are damaged.

Though they vary in size, vessels are larger than other hardwood cells and are often quite visible (in oak and walnut, for example) as holes when a crosscut log is viewed from the end. Visible end-grain holes are referred to as pores (board ends stain darker than the faces because the pores hold more stain than the less porous faces). Since all hardwoods have vessels, they are often described as porous, while softwoods, which do not, are nonporous. Since hardwoods do not secrete resin like softwoods, air-drying is not as much a problem as it is with softwoods.

Among others, oak and walnut are considered ring porous because most of the larger pores are in the earlywood layer of the annual ring. By contrast, lumber from maple and poplar are considered diffuse porous because their smaller pores are spread out over both the earlywood and the latewood layers. Since more stain accumulates in the larger pores, the color of the stain will be more pronounced in a ring porous wood like oak than in a ring diffuse wood like maple.

In both softwoods and hardwoods, cell assemblages that run horizontally from the pith to the bark are known as rays. They conduct sap across the grain of the vertical cells. Rays are very visible at the ends of oak and sycamore logs, two trees that are often sawn in a way that emphasizes the visible presence of the rays (a method known as quarter sawing).

Unless I need wide boards, I quarter saw all of the oak I acquire. I like both the appearance of the grain and its stability once dried. Ray cells are much narrower in softwoods and are not visible to the naked eye. Quarter-sawn softwoods are stable but lack the visible ray fleck of oak or sycamore.

In some hardwoods, such as white oak, there is an abundance of bubble-like growths in the cell lumens called tyloses. Tyloses partially or even completely block the direct flow of sap. Tyloses in both American and French white oak make both suitable for wine and whiskey barrels since they make the barrel liquid tight (barrels made from red oak will not work because, with few exceptions, tyloses do not form in red oak). Tyloses do make white oak slower and somewhat more difficult to dry and to treat with preservatives.

Water Content

Though the shapes, sizes, functions and positions of cells vary throughout a tree, they all have walls that create an enclosed cavity that can hold sap (mostly water), air, or living protoplasm. Water is considered bound in the cell walls and free in the cell cavity. As lumber dries, free water in the cavity is the first to go. Once the free water is gone, the boards have reached what is called the fiber saturation point. Bound water is removed from the cell walls by placing the boards in a drier setting; that is, where the relative humidity of the surrounding air is consistently low enough to pull water from the cell walls. This is what kilns do. Wood drying is discussed in the next chapter.

Additional Information

For more detailed information on the structure and growth of trees from the perspective of woodworking read *Wood Handbook, Wood as an Engineering Material*[15] prepared by the USDA Forest Service's Forest Products Laboratory, R. Bruce Hoadley's *Understanding Wood* (revised),[16] and Brayton F. Wilson's *The Growing Tree.*[17] All three are excellent references. For a detailed description of the reproduction, growth and structure of one very familiar genus, the oaks, read Glenn Keator's *The Life of an Oak.*[18]

Recognizing Quality Trees

The quality of lumber that can be sawn from urban trees depends on the species, the size of the tree, where and how it grew, whether it has been damaged by natural or human action, and what foreign material might be embedded in it. The odds of getting good quality lumber are highest when the most valuable trees have grown on level ground, reached maturity without significant damage from wind, lightning, disease, insect infestation, or construction, and contain no concrete, metal, or anything else that the trees themselves did not grow.

Species

Ed Lempicki and Ed Cesa (from the New Jersey municipal tree project described in the last chapter) assembled the following list of species by their general commercial worth, from best to poor. Log sizes are similarly ranked.[19]

1. General Species desirability

Best: Walnut, butternut, ash, oaks (except pin oak), cherry, and paulownia

Good: Hard (sugar, black, and rock) maple, elms, most fruitwoods, basswood, sycamore, cedar, yellow poplar

Fair: White pine and other softwoods, mulberry, osage orange, persimmon, beech, soft maple, and red and silver birch

Poor: Gum, ailanthus, pin oak, aspen (also called poplar or popple), and cottonwood.

2. General size requirements

Best: 16 inches or larger diameter at small end of log 10 feet or longer in length

Good: 14 inches or larger diameter at small end of log 8 feet or longer in length

Fair: 12 inches or larger diameter at small end of log 8 feet or longer in length

Poor: small in length and diameter; or large in length and diameter with many knots, branches, holes, rot, or cracks; or with large or numerous metal objects.

Roughly half the calls I receive about oaks are from homeowners wanting to know if what turns out to be a pin oak has any value. That proportion of calls about pin oaks is not surprising since it is the most popular urban oak in the nation.[20] It is disease resistant, tolerates air pollution, and adapts well to often-sparse urban soil. It is a common tree in New York's Central Park. Unfortunately, pin oak lumber has limited value because of the small but numerous holes throughout the lumber. These are small loose knots that fall out once the lumber begins drying. The stack of pin oak lumber I examined looked as if someone had drilled small holes in a random pattern from one end of the boards to the other. Though I have never seen anything made from this species, an imaginative woodworker might find a way to use the holes as a feature of a box, turned bowl, or even a piece of furniture. (Who would have thought that pecky cypress would be used as rustic paneling?) It could be used outdoors for parking lot blocks or park trail steps.

Good osage orange logs are hard to find because the trunk of the tree is often short. Even when tall, the trunk twists as it grows. The lumber cut from osage has a prominent and irregular grain that is dramatically highlighted by a combination of bright shades of yellow and gold. Unfinished, the colors fade. However, a polyurethane finish can fix the bright shades. The wood can be difficult to work; thickness planers tend to tear out small chunks. Nevertheless, from my own experience with osage, I believe it is worth the effort. Native Americans used it for bows, its yellow color was extracted for dye, and in times past it was planted as a hedgerow. Livestock will eat the green softball-size fruit called horse apples.

Sycamore and oak are the only two eastern hardwoods which, when quarter sawn, have prominent ray fleck, small irregularly-shaped smooth spots of grain that are a feature of fine Stickley and Arts and Crafts furniture. Because of the stress it undergoes when drying, the best way to get usable lumber from sycamore is to quarter saw it.

Although hickory and the closely related pecan are not on the list, they are, in my judgment, woods worth having. Both grow throughout most of the country, from the mid-western to the southern states and west to Oklahoma and Texas. Because it is hard and resilient, hickory makes good handles for tools, like sledgehammers, that take a lot of abuse. It is also ideal for bentwood furniture. That it is being used to make kitchen cabinets indicates growing acceptance as well. As I pointed out in the last chapter, I have worked with pecan and highly recommend it for furniture.

Further west in California urban areas, there are more than two dozen species of valuable urban trees including acacia, eucalyptus, camphor, cypress, California black walnut, tanoak, black oak, madrone, redwood, monterey pine and douglas fir. A complete list of valuable hardwoods and softwoods by commercial, common, and botanical names is given in Appendix A.

For more detailed descriptions of both domestic and imported wood, see *The Encyclopedia of Wood*[21], *Woods of the World*[22], Wood Profiles at www.woodmagazine.com and the Species Guide at www.hardwood.org.

Again, you can also rely on local and state forestry officials and tree service arborists for information on types of local trees.

Reaction Wood

When a tree grows at an angle, as it might on a hillside where there has been ground slippage, or near a creek or river where the ground has been eroded, the pith is not centered in the tree. Instead, it is off-center and the surrounding growth rings are not round but are more oblong in shape. Off-centered growth creates reaction wood: the trunk or limb is reacting to the forces tilting it, by trying to grow upright. Reaction wood also forms in limbs that do not grow upright. More specifically, when it occurs in softwoods, it is called compression wood. In hardwoods, it is called tension wood. Compression wood in softwoods is created on the underside of the leaning part of the trunk: for example, if the tree is on a slope and is leaning or curved downhill then the compression wood will be on the downhill side of the trunk. In a leaning or curved hardwood, tension wood occurs on the opposite side of the leaning angle. Imagine yourself standing straight with your arms also straight above your head and your hands clasped together. As a kind of yoga exercise, as you lean to the right, you can feel the pull or tension in the muscles on your left side. On your right side, you can feel the skin and muscles being squeezed or compressed. Now, on your right side, you know what compression wood in softwood feels like and, on your left side, what tension wood is like for the hardwood.

Reaction wood can be very difficult to cut, dry, and finish. Lumber being cut from a log can pinch the blade when the sawing itself releases the built-up tensions in the wood. Having cut reaction wood both from stems and limbs, I can attest to the power of a board, suddenly freed of its stress, to abruptly stop the band saw blade in its tracks. Only a crowbar and, ultimately, a chain saw, allowed me to back the blade out of a cut that had squeezed shut tight enough to trap the blade between the board and the log.

Because of differential shrinkage during drying, reaction wood does not behave well when either air-dried or dried in a kiln. In addition to the usual ways lumber distorts itself when drying (cupping and cracking, for example), boards cut from reaction wood can shrink lengthwise as well. Normally, boards shrink mostly across the grain, usually the faces of the board, and very little in length. Compression wood can shrink up to ten times as much as normal wood, and tension wood up to five times the norm. Since this shrinkage is not uniform, further damage is done to the boards as they dry.[23]

Reaction wood might not take finishes, especially stains, very well. Because the wood cells are distorted, some parts of a board may absorb far more stain and finish than others, producing an undesirable extreme of color and shade in the finished piece. In some cases, the surface of the board remains fuzzy, even after fine sanding, so that both the feel and the appearance of the piece is compromised. In all, there has to be a compelling reason to harvest an urban tree that is known or suspected to contain reaction wood.

Wind-Damaged Trees

Assuming the tree has grown upright and contains no reaction wood, there are other problems that occur during growth that need to be considered. Among natural forces, high wind is the most frequent cause of damage to urban trees.[24] Breaking limbs can damage the trunk of the tree, reducing the amount of lumber that can be cut from it. Trees felled by high winds are often damaged by ground impact; they either split along their length, crack or break crosswise. Ground impact can also cause separation between the growth rings, a condition referred to as shake or ring shake. While lumber can be cut from trees felled by high winds, clearly this is not the ideal way for them to be harvested and the resulting yield will be considerably less.

Some trees subjected to high winds remain standing and, from the outside, appear undamaged. However, separation has occurred between the growth rings, a condition called wind-shake. Often, this damage is not noticed until the logs are being sawed and, sometimes, not until the lumber begins to dry. In my experience, this wood is generally not usable as lumber.

Other Natural Disasters

Uprooted hardwoods and softwoods should be quickly harvested before insects attack the downed or weakened trees.[25] The same applies to trees with major wounds and limb loss. Bent and broken hardwoods will survive longer than softwoods but are still vulnerable to decay brought on by fungi. Pines that have lost their tops and have no live limbs will soon die. Both should be felled soon.

Often major natural disasters, such as the eruption of Mount St. Helens in 1980 or Hurricane Iniki that struck the Hawaiian island of Kauai in 1992, can knock down so many trees at once that quickly reclaiming them is a major challenge. After all, the first priorities in such disasters are rescuing victims, providing medical treatment for the injured, and restoring power, communications, and transportation. Storms produce an enormous amount of debris that must be cleared quickly, especially in urban areas. Though salvaging trees is certainly not among the highest priorities in these emergencies, as debris is cleared an effort can be made to divert potentially usable trees to a separate site where they can be sawed at a later time. And, diverting trees would take some of the immediate pressure off of the landfills.

Parenthetically, over a two-year period, the Weyerhaeuser Company hauled out about 600 truckloads of logs daily from company property located within the blast zone of Mount St. Helens. Trees flattened by the tremendous force of the eruption eventually yielded about 850 million board feet of softwood lumber (mainly fir, cedar, and hemlock), enough to build 85,000 new homes, according to Weyerhaeuser.[26]

As for Kauai, we left the island on September 7, 1992, four days before Hurricane Iniki struck with such devastating force. We returned about two years later, in part to see if, and how, trees blown down by the hurricane had been used. Iniki damaged or destroyed about 80 percent of the homes and left the island without power for 40 days. We learned that the pressure of recovering from such massive damage, plus uncertainty about who was responsible for the hundreds of trees blown down, resulted in the loss of most of the wood. While some trees were retrieved, we were told that most rotted. This includes koa, one of the finest and most expensive of American hardwoods, and widely recognized as a symbol of the state.

Lightning also damages trees, fatally if the strike's electrical energy surges through the cambium layer of cells. Trees generally survive strikes that move through the bark. Lightning tends to leave a noticeable path of stripped or damaged bark down the side of the stem. Outside this strip, the remaining wood should be usable.

Insect, Fungal, and Bacterial Damage

There are a large number of insects and microbes (mainly fungal and bacterial) that can damage and kill trees. The Asian longhorned beetle is an example of a recent arrival in the U.S., possibly from China, that poses a serious threat to maples, horse chestnuts, poplars, willows, elms, and black locust.[27] By continuously boring into and feeding on the cambium cells (recall that these are the cells that produce more of their own as well as bark and sapwood cells), the larvae work their way around the tree. Once girdled, the tree dies. Since there is as yet no known biological or chemical defense against this insect, the only recourse is to cut and destroy infected trees.

First discovered in 1995, Sudden Oak Death syndrome is responsible for the deaths of tens of thousands of tanoaks, coast live oaks, black oaks, Shreve's oak, and madrone from Monterey to Mendocino counties in California.[28] Caused by a newly discovered pathogen, *Phytophthora ramorum*, little is known of how it spreads from tree to tree, whether it can jump to redwood and douglas fir, or whether it could spread beyond California to the eastern United States (northern red and pin oaks in the east are possibly at risk).

These are two cases among more, where the best action is to cut and dispose of the infected trees in a way that minimizes the spread of formidable insects and diseases.

While usable lumber could be cut from trees killed by the longhorned beetle and the California oak fungus, the trees should be destroyed instead. The potential cost from spreading infestation and disease is likely to be far greater than the loss of the lumber.

Standing dead or dying trees may be harboring one or more fungi. The most obvious sign of advanced fungal rot is the presence of fruiting bodies or conks growing on tree trunks and limbs (see Figure 19 below). By the time these growths appear on the tree, the wood inside, including the more rot-resistant heartwood, has already been invaded and damaged. Trees in this stage are not likely to yield much usable lumber. Other fungal infections, not so apparent from the outside, may simply stain the sapwood a bluish color without actually damaging it. This wood can be put to use where surface appearance is not important.

In the early stages of some fungal rot, thin dark lines and shading are produced in the wood that are attractive in panels in boxes and cabinets and turned bowls. The visual effect is referred to as spalting and is commonly seen in maple and oak. The dark lines create patterns that look to me like abstract pen and ink drawings. See Figures 20 and 21. Be warned, however, that dust from cutting and sanding spalted wood may provoke an allergic reaction. After cutting several spalted hard maple

FIGURE 19
Conk growing on side of locust tree. Wood inside is rotten.
Photo by Sam Sherrill

logs, I had a reaction that nearly closed my throat. While I like the visual effect of spalting, I have given up cutting logs that show signs of fungal infection. I cut only very small amounts for boxes and only when I'm wearing a truly effective dust mask.

As I pointed out above, separation between the growth rings, referred to as shake, can be caused by ground impact or wind. Shake can also be caused by enzymes secreted by bacteria that have infected the tree. At the growth rings, the enzymes damage the wood cell walls in a way that leaves the walls largely intact while causing the cells to break away from their neighbors, thus causing separation between the rings. The water in the cells leaks into the air spaces around the cells creating a condition referred to as wet-wood (or bacterial wetwood). Wetwood is a major source of economic loss of red oak

FIGURE 20
Spalted maple panel in a walnut box I made for a friend.
Photo by Sam Sherrill

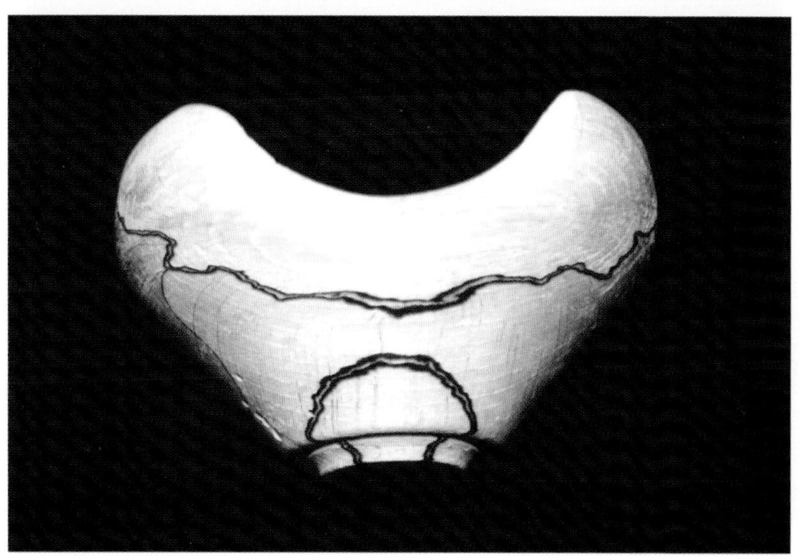

FIGURE 21
Occasionally the spalted pattern looks as if it were drawn by pen and hand, like the outline of a mushroom on the bottom of this bowl.
Photo by Sam Sherrill

lumber.[29] An urban tree with this bacterial infection, once felled, will have a distinct rancid odor and be saturated with moisture, so much so that in bucking and limbing the logger may be sprayed by the foul-smelling liquid. Wetwood does not dry well: cracks open up deep in the lumber and the wood tends to warp. In addition, the kiln can be corroded by acidic fumes escaping from the drying wood. These same fumes can make drying wetwood indoors, in a basement workshop or other area connected to living and working spaces, unpleasant if not hazardous.

Heavy Metal and Concrete

The main reason many sawmill owners emphatically reject urban trees is their fear of the metal, concrete, and other hard objects that may be embedded in the wood. For these mills, even small metal objects—nails, tacks, and wire fencing—can damage the saw blade. Most circular saw and band saw mill blades will cut through small soft metal objects (such as a lead bullet) without damage to the blade, especially if the object has thoroughly rusted. Carbide-tipped blades are even less vulnerable than standard steel blades. Some band mills have a debarker (also known as a mud saw), a small vertically positioned circular saw with a carbide-tipped blade that cuts a narrow path just ahead of the blade (there are debarker attachments for chain saws as

FIGURE 22
Steel eyebolt embedded in a log. Steel cable attached to the eyebolt is used to support limbs and trunks that might otherwise fail under their own weight or wind pressure. The eyebolt is thick enough to be a blade killer.
Photo by Sam Sherrill

well). The debarker hits the object first and either cuts it or alerts the operator that something is there.

A lot of strange things end up in trees (look at Figure 17 again). Hall reports finding horseshoes, musket balls, old glass insulators, and a full can of beer![30] Small car parts may emerge where shade tree mechanics have worked. An employee of the Cincinnati Parks Department told me he once found a spoon deep inside a tree. There is even one report of a human skeleton being found. The skeleton is thought to be the remains of a French soldier, possibly wounded in battle, who hid in the hollow of a tree sometime during the French and Indian War. His hiding place became his tomb.[31]

While small objects might not pose much of a hazard, large metal spikes, lag bolts, and eyebolts used to cable limbs together can shatter circular saw blades and break band mill blades. The eyebolt shown in Figure 22 would be one such hazard. Shattering and breaking blades are dangerous to sawyers. In addition, the blades themselves are expensive to repair or replace, and money is lost while the mill is idle.

Logs can be visually inspected for embedded or protruding objects and for dark stains produced by ferrous metals, especially in oaks. Most metal driven into trees, especially nails, turns up about five to six feet above the ground where people can easily post a flyer for a lost pet or garage sale. At about the same height, you may find eyebolts used for clotheslines and hammocks. At eye level in very old trees, you may even find musket balls and arrowheads. Portable metal detectors can also be used for ferrous metals and there are scanners used in large-scale production settings that will detect nonferrous metals as well. Metal and other objects should be marked. I use a liberal amount of white spray paint on the objects. I paint each end of the log with red paint so that, as logs accumulate, I do not lose track of those needing further cutting or trimming. I try to cut around the object in a way that minimizes the loss of wood. However, if I'm in doubt, I either trim the log or reject it.

Concrete and glass objects produce no telltale stains nor can they be found using detectors. Recall from Chapter One, that Lynn Erickson, a former log buyer for Minnesota Hardwoods, recommends avoiding urban trees with two or more trunks because they are more likely to have been cabled together. He suggests avoiding trees that have grown close to homes because they typically contain more metal. And, he avoids hollow trees as they are the ones most likely to contain concrete. These recommendations help reduce, but do not eliminate, the likelihood of hitting an undetected and undetectable object. Unfortunately, the first sign that something has been missed is the sound of the blade hitting it. Transportable band saw mills that use relatively inexpensive blades are best suited for urban trees. At least an expensive blade will not be lost and the machine will not be idle any longer than the time needed to disengage and change the blade.

Harvesting Mechanics: 4 Ways to Fell Urban Trees

Trees can be felled four ways: manually using just a chain saw; by a chain saw along with rigging and additional equipment such as a crane; by large earth moving equipment, or by commercial logging machinery. The first two are appropriate for more densely-populated metropolitan centers and urban neighborhoods. Usually, just a single tree, or at most a few trees, are taken down on any one site. The second two methods, both of which require heavy equipment, are appropriate for large tracts of land being cleared of many trees to make way for shopping centers, commercial and industrial parks, residential development, or new roads and expanding beltways. Large-scale clearing usually takes place in metropolitan counties where urban expansion overruns wooded areas, forests and farmland. These are the less urbanized parts of metropolitan counties, where large numbers of trees can still be found.

Individual property owners are likely to fell one or a few trees using just a chain saw. If you do this yourself, the tree should be felled first and then limbed and bucked on the ground. Described below are safe methods of cutting with a chain saw that preserve the value of the log, and safety measures that should be followed when using a chain saw. You should read all of this before deciding whether to bring down trees yourself.

Limbing standing trees is far too risky for amateurs and should be left only to tree service companies with the right equipment and crews who can safely do this kind of work. The second cautionary story below describes what happened to someone who ran the risk.

Urban arborists and tree service companies usually take down trees, one or a few at a time, by limbing them while they are standing and then bucking them in sections, each of which is lowered or dropped to the ground. Because of training and experience, their crews are well prepared to use chain saws aloft, rigging for climbing and for lowering limbs and, on occasion, cranes to extract large limbs and logs from tight spots on urban streets and properties.[32]

Unlike the work of loggers, 75 percent of the work done by these crews is cutting and trimming while aloft in trees. Felling trees accounts for only about 5 percent of what they do.[33] Even so, they are in a unique position to safely take down a tree in a way that preserves its value as a saw log. They are also qualified to judge the condition of what they are taking down, an assessment that helps owners decide whether the trees will later yield good quality lumber. Indeed, arborists and tree service companies, along with urban parks and public utilities crews who do the same work with the same equipment, are key components in any organized effort to harvest urban timber.

For large urban projects, some developers reduce clearing costs by selling the logs to sawmills (rising timber prices further encourage this practice).[34] All the felling could be done by crews equipped with chain saws. However, large track-laying front-loaders used for earth moving and excavation can efficiently fell and move large numbers of trees. As described in the next paragraph, loaders can quickly fell trees in a way that preserves their value as saw logs (because of the limited movement of the blade, bulldozers are not well suited for this). Since the loader is already present for site preparation, the marginal cost of felling should not add much to the total cost of its use. Damage during felling is not as important when the logs are being sold for pulp, other than how it might complicate loading and hauling. In that case, any machine that is already on-site, and capable of knocking down trees, can be used.

As pictured in Figure 23, the loader operator can knock over a large tree by excavating around its base enough to partially expose the roots. The front edge of the bucket is then pushed beneath the partially exposed roots and raised, partially uprooting the tree. Next, the operator backs up enough so that the bucket is clear of the roots. The bucket is then raised and rotated back toward the cab so that the bottom of it is facing forward and the bucket teeth are facing upward. The loader is then slowly driven, bucket first, against the trunk at a height of 10 to 15 feet. This works against all but the very largest trees (those must be felled by chain saw and the root system excavated or chipped out later). The finesse with which the operator pushes the tree over determines whether

FIGURE 23
The operator of a tracked front-loader has partially uprooted a maple tree. The final push felled the tree down without damaging the trunk too much.
(Photos by Sam Sherrill

FIGURE 24
Less damage would have been done if the bucket had been rotated back so that the bottom of it would have been pressed against the trunk instead of the teeth.

the trunk will yield a saw log or simply provide splintered and broken wood suitable only for a tub grinder (which is what happens when the only goal is to quickly clear the land). The more the tree is uprooted and lifted to an angle, the less damage the bucket will do when the loader gives the tree a final push. The teeth on the front edge of the bucket will damage the bark and probably the sapwood, as pictured in Figure 24. However, damage can be minimized when the operator rotates the bucket back toward the cab as far as it will go and pushes the flat underside of the bucket against the trunk. Spreading the pressure on the underside of the bucket will not damage the tree like the teeth will. Uprooting and toppling a tree by using only the raised bucket, teeth first, can substantially damage the trunk

FIGURE 25
A John Deere feller buncher cutting a tree at ground level.
(Used by permission of Deere & Co.)

FIGURE 26
A John Deere skidder grappling and hauling felled trees.

at a depth and position where the best lumber is located. After the tree has been felled, the operator can use the corner teeth of the bucket to snag the root and, with the bucket raised several feet off of the ground, drag it to the edge of the site where it can be bucked and limbed.

Uprooting trees not only creates an inexpensive opportunity to preserve their value as lumber, in most cases it is faster and less expensive than felling trees by chain saw and then digging out their roots later with a loader or bulldozer. Uprooting also preserves highly figured and highly valued wood in the larger roots of such trees as walnut that is prized by woodworkers for turning bowls, making gunstocks, decorative boxes, and rustic furniture.

Harvesting timber with a loader depends on having the approval of the developer and the cooperation of the contractor, as well as precise communication with the loader operator. In my experience, being present during the clearing operation to describe what needs to be done and to answer questions, helps ensure that the trees are felled with minimal damage.

Modern commercial logging companies use machinery designed for large-scale felling and removal. Two important examples are the feller buncher, shown in Figure 25, and the log skidder, shown in Figure 26. The feller grasps the tree and cuts it near ground level. The skidder gathers felled trees in a large grapple hook and hauls them to a landing area where they are loaded on trucks and hauled to sawmills. While such machines might be appropriate for very large land clearing operations at the edges of metropolitan areas, they are too expensive for small operations, and clearly inappropriate for the typical urban streets, yards and parks. Aside from their sheer size and weight, a tracking-laying machine like the feller buncher in Figure 25 would tear up asphalt and concrete pavement. In addition, the total cost of using such equipment on all but the largest clearing projects would be much greater than the marginal cost of using a loader, even though the loader is not as well suited to felling.

Few readers of this book will be using heavy equipment to fell trees. While you should be aware of these options, instructions are not necessary since you will not be using arborists' rigging to climb and cut trees, nor will you be operating cranes, loaders, or feller bunchers. **However, many of you either use, or might consider using, a chain saw to fell, limb and buck trees on your own. You should not continue or even start doing so until you have read the rest of this chapter on felling and cutting trees and on safe chain saw use.**

Felling Trees with Chain Saws: Amateurs Beware

Two Cautionary Stories

Several years ago, I was standing next to a large cherry tree watching it being cut by someone with many years of experience felling urban trees. The tree was between forty and fifty feet high and about thirty inches in diameter. He cut a conventional notch facing the direction he wanted the tree to fall. After removing the wedge out of the notch, he then made a back cut on the opposite side. However, instead of falling directly away from the back cut, as it was supposed to, it fell ninety degrees from the intended direction. Fortunately for me and for several others close by, the tree fell away from where we were standing. We were also lucky that the tree did not kickback on the stump as it fell (kickback occurs when the tree quickly slides or jumps back over the stump in the opposite direction of the fall). Had it kicked back in our direction, one or more of us would have been struck by the butt end of the trunk and seriously injured, or worse.

For free, I learned several important lessons that day. First, I have no interest in felling trees. I do not have the temperament, training, experience or equipment, nor do I have the necessary liability insurance. As the OSHA Logging Advisor in Figure 27 points out, **felling trees is the most dangerous part of commercial logging. I believe this can be just as true for urban logging as well.** Even if no one is injured, there is always the possibility that the tree, or some part of it, will damage buildings, vehicles, yards, gardens, or power lines. Second, there is a degree of uncertainty about where trees and limbs will fall, even when professionals cut them. For amateurs, we must assume that the uncertainty is much greater—much too great to risk. Third, if you are not cutting, then you have no business being close to the tree being felled. All of us should have been stand-

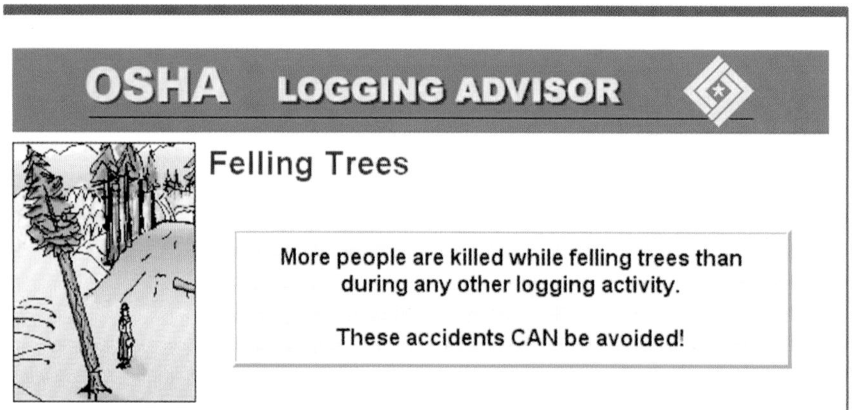

OSHA LOGGING ADVISOR

Felling Trees

More people are killed while felling trees than during any other logging activity.

These accidents CAN be avoided!

FIGURE 27
Something to think about before felling a tree.
OSHA Logging Advisor

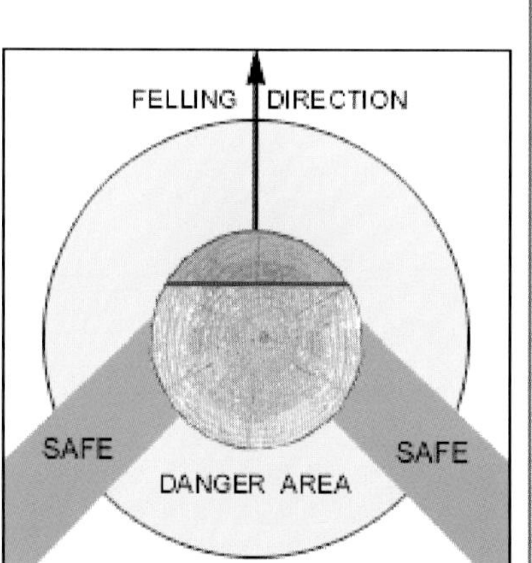

FIGURE 28
As shown by the green SAFE areas above, you should know where the safe escape paths are as well as the DANGER AREAS.
OSHA Logging Advisor

ing at least two tree lengths away. Finally, if you must be within the potential fall zone, then a clear escape path should be designated before cutting begins.

Figure 28 from OSHA indicates safe retreat paths, assuming the tree falls in the intended direction. Make sure the path is clear of obstructions so that you can run without colliding with or tripping over something. Just like fleeing a fire in your home or a hotel room, you will not have time to ponder which way to go if things should suddenly go very wrong.

By sheer luck, my own experience is a cautionary tale that ended well. Not all stories do. In the summer of 1982, a 21-year-old college student, who operated a lawn service during the summer, climbed a seventy-five-foot hackberry in one of Cincinnati's upscale neighborhoods.[35] He had been hired to trim limbs that blocked the homeowner's view of the nearby Ohio River valley. A large limb, which he had cut, did not fall away from the tree but instead fell back into the crotch of the tree where he was positioned. The limb pinned him upside down by his left arm and leg. Fortunately, the owner of Madison Tree Service, Ed Butcher, and his son Jack, working nearby, heard about the accident and went to help. Ed's company was one of the few in Cincinnati at the time that had training in rescuing people trapped in trees. Jack Butcher worked for over two hours to free the trapped man. During that time, medics administered saline and pain medication. After enough of the limb had been cut away, Jack and the medics were able to free the man and rush him to waiting surgeons at a local hospital. Though he survived the ordeal, his left arm and leg had to be amputated.

Worth keeping in mind is that in the language of logging, a "widow maker" is any loose overhead debris, such as limbs or treetops, which can fall unexpectedly.[36]

The Task of Felling: Who and How

Please note: if you want to take down a tree on your own property, especially near buildings, power lines, or streets, you should first receive professional instruction and supervised practice. Without training and experience, you are very likely to injure yourself or someone else, or damage property or power lines. If you do not have or intend to get training, you should leave this work to professionals. This advice goes double for trees on someone else's property.[37]

If the tree is in an open space well away from buildings, power lines, streets, gardens, adjacent property, or anything else of value, you might decide to take it down yourself. If this is your choice, you should fell the tree first then limb and buck it on the ground. **Without training and a professional crew to help, you should not attempt to limb a standing tree (read again the second cautionary story above).** Also, keep in mind that when the entire tree is dropped, its protruding limbs will gouge the ground and flatten the grass beneath where it lands, so you will need to devote some additional time to repairing the damage.

I have felled several trees but, for the reasons I listed above, I do not do it any more. **I hire and I recommend hiring pro-**

fessional fallers. While you might save on the expense of paying someone, you could end up paying a lot more if something goes wrong. Once the tree starts to fall, there is no turning back, no opportunity to reconsider your choices, and no way to predict what is going to happen next—and happen very quickly.

However, even if you do not fell the tree yourself, you can and should be a knowledgeable observer (from a safe distance). You should be able to ask critical questions about how the tree will be felled and evaluate from the answers whether you feel comfortable with the felling method being proposed and that the method will yield the best saw logs. This is the point where you would indicate that you want the tree for its lumber, arrange to have it cut appropriately, and either hauled to a sawmill or moved to an accessible place, on your property, for the mill.

Reputable tree service companies have trained crews who are concerned about the safety of people and property. These companies should be able to provide proof that they are covered by liability insurance, are enrolled in your state's worker's compensation program, and should be able provide you with a list of customers as references. Of course, they should be willing and able to fully answer all of your questions.

There are freelancers with chain saws and pick-up trucks who present themselves as professionals, but are not. There are considerable risks associated with hiring them: at the very least, they may not be as knowledgeable as they claim to be (worse yet, they may not know what it is they do not know). Even if they do have adequate skills, they may not own or be willing to rent adequate heavy equipment. They should be able to present up-to-date evidence of liability coverage and of enrollment in your state's worker's compensation program. Be suspicious if the insurance or worker's compensation documents are "at home" or "in the other truck". I would not proceed until I have seen all of the documentation, determined that it is current, and have been given a list of references. Property owners can be held legally responsible for injuries and damage caused by uninsured freelancers.

If you decide to do it yourself, the following are essential points to review before and right after taking a tree down:[38]

1. Do not work alone. Have at least one other person present in case of an emergency.

   Professional loggers working alone have died from otherwise survivable injuries because no one was present to call for emergency medical assistance.
2. This should not be treated as six-pack Saturday afternoon entertainment: no one involved should be drinking or taking drugs while using a chain saw, whether felling, bucking, or limbing. In addition, check medications for warnings about drowsiness.

   No participant should be working while impaired either chemically or by a casual attitude.
3. Conduct a "walk through" to assess conditions and terrain. Make written notes about potential hazards and keep the notes available for reference when cutting starts.

4. Develop a felling plan that specifies what is to be done, the sequence of tasks, and who will be doing them.
5. Be aware of weather conditions, especially changes in conditions that bring on high winds, lightning, heavy rain, hail, or heavy snow. Even if the wind is not high, watch its direction and keep in mind that a sudden gust could push the tree away from the direction you want it to fall.
6. Make sure that everyone involved or observing, as well as everything of value, is at least two tree lengths from the felling site: that is, both everyone and everything should be outside the yellow danger zone illustrated in Figure 28 above.
7. Clear away brush, limbs, vines, nearby small trees, and other obstructions that might interfere with the tree as it falls.
8. As illustrated in Figure 28, establish and clear escape paths.
9. Check the tree to see which way it is leaning. Also check for broken limbs still clinging to the tree or fallen limbs lodged among the other limbs. Look for significant crooks and defects in the trunk. Look for beehives, vines that connect to adjoining trees, and steel cabling in a single tree that ties limbs together or two or more trunks together. Vines and cabling could affect where the tree lands and how far debris will be thrown when it hits the ground. Angry bees will send everyone running, especially those allergic to bee venom.
10. Check nearby trees for dead and lodged limbs that might be dislodged by the tree being taken down. Determine whether any of these trees might interfere with the falling tree; that is, catch or hang it.
11. Before you begin cutting make sure everyone is out of the danger zone and understands that, once you start, they are to stay out. Tree felling seems to draw spectators who do not know unless told, sometimes with emphasis, that they must stay well away from the work area.
12. After the tree is on the ground, carefully inspect it for hanging limbs, broken tops, or both, that are not on the ground. Look for limbs under back or sideways pressure which, when cut, could lash out suddenly and with tremendous even deadly force. Be aware that if the trunk itself is suspended above the ground on one or more limbs, then it could roll toward you when the limbs are cut. Finally, look for smaller trees that might have been caught and bent by the felled tree. Called spring poles, these too can cause injury when the tension they are under is released and they spring back to their original position.

These dozen recommendations might seem like a lot to do just to bring down a single tree. However, checking each point will not take all that much time initially and could save you a lot of time, money, and grief later on. Give this list the same careful and focused attention you would want the pilots to give to the pre-flight check before your plane takes off or the preparations you would want your surgeon to review before you undergo surgery.

Under the following circumstances, you should not even consider taking a tree down on your own but should either wait until later, or call professionals instead:

1. There is no one to work with you or be present as an observer.
2. You, or other essential participants, are impaired by alcohol, drugs or medications.
3. The tree is either on a steep hillside or is otherwise leaning markedly in one direction, especially toward power lines, buildings, streets, a fence, valued vegetation, or an adjacent tree. (Leaning trees may not be worth harvesting for furniture-grade lumber because they yield stress wood that neither dries nor works well.)
4. When the tree, toppled by wind or lightning, did not make it all the way to the ground but got caught in another tree, on a power line, or is lying on a building. If the tree is not flat on the ground, do not touch it.
5. When the tree has more than one main trunk or stem, especially when the main trunk divides into two or more large stems well above ground level.
6. A tree whose limbs or multiple stems have been cabled together.
7. The tree has a large and lop-sided crown that could unpredictably shift the direction of its fall despite how it has been cut. These trees often need

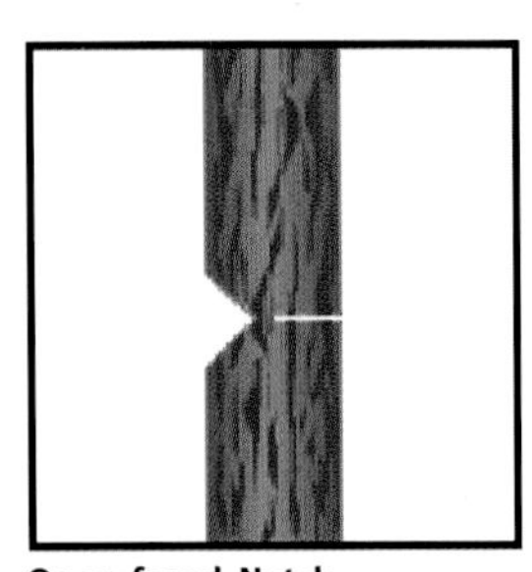

Open-faced Notch

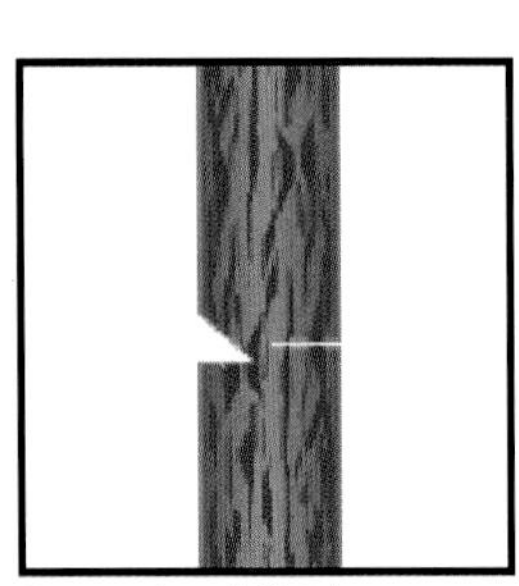

Conventional Notch

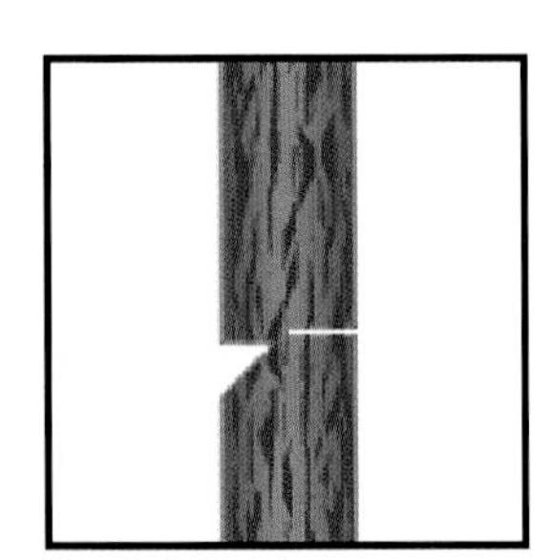

Humbolt Notch

FIGURE 29. Three types of notches and accompanying back cuts recommend by OSHA.
OSHA Logging Advisor

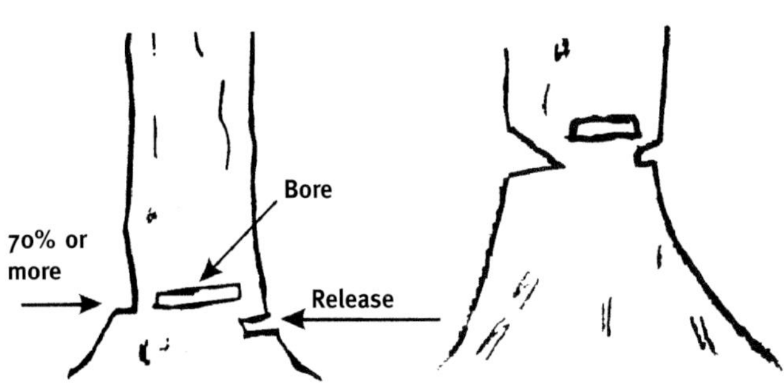

Open-faced Notch
Bore cut at same level as notch

Conventional Notch
Bore cut slightly above notch

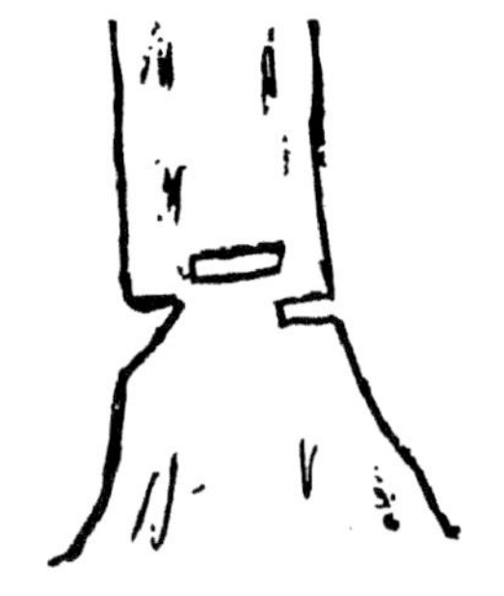

Humbolt Notch
Bore cut slightly above notch

FIGURE 30. Three notch cuts and accompanying bore cuts recommended by FISTA.
Arborists' Chain Saw Safety Training Guide

to be topped while still standing and then pulled in the desired direction by cable and winch.

8. If the tree you cut hangs up on an adjacent tree or stalls on the stump (see Figure 33), cordon off the new danger zone and call for professional help. Hanging or stalled trees are extremely dangerous and can only be brought down by professionals who know exactly what to do and have the machinery to do it. At this point, recognize that you are in over your head and that you should not attempt to complete the felling.
9. Finally, trust your own survival instinct: if you feel uncomfortable about dropping a tree, even if you cannot quite put your finger on why, trust the feeling and either wait until later or call a professional.

Manual Felling with a Chain Saw

I have chosen the felling methods recommended for loggers by OSHA and by FISTA, the Forest Industry Safety and Training Alliance. Both provide specific instructions and emphasize safety above all else. **Even though I have selected what I consider to be the safest felling methods, you must consider what follows as informative, not instructional.** You cannot take these or any set of rules from the page to the tree: there must be instruction and experience between the two. Once the tree starts to fall, you cannot change your mind or be caught still reading the directions.

Following OSHA's instructions (shown in Figure 29), three cuts are made in the trunk of the tree. Starting on the side of the trunk facing the landing zone, one of three types of notch cuts is made.[39] The options are open-faced, conventional, and Humbolt notches (there are other specialized cuts for trees in difficult settings but we will review just these three basic ones).

According to OSHA, the open-faced notch is the safest of the three because the tree is more likely to fall where intended and there is less chance of kickback. The conventional and Humbolt notches are more frequently used because they are most familiar to loggers and tree service companies. As illustrated below, the open-faced cut forms roughly a 70-degree opening on the side of the tree trunk. The conventional notch consists of a straight cut into the trunk followed by a 45-degree cut down from the top. The Humbolt notch is the reverse of the conventional one, in that the angular cut is made up, from below the straight cut.

For all three notches, the first two cuts are made into the trunk to the depth of about one-third the tree's diameter. So, for a tree that is two feet in diameter, the first and second cuts should be about eight inches in depth. When the two cuts meet precisely, the wedges will fall right out of the notches.

Next, a third or back cut is made from the opposite side, as illustrated in Figure 29. The back cut does not connect with the notch and is slightly offset from the horizontal cuts of the conventional and Humbolt notches. The back cut for the open-face notch is at the level where the two cuts meet. For all three, the wood left between the front and back cuts serves as a hinge that helps prevent kickback and directs the tree's fall toward the landing zone. If the tree does not fall after the back cut is made, then a plastic or iron wedge is carefully hammered into the back cut. The tree should start tilting as the wedge is driven further into the cut. At some point, the tree will fall. As it does, the logger should shut the saw off and, as illustrated in Figure 28, quickly move along an escape path to safety.

FISTA recommends and teaches another method of back cutting that is gaining wider acceptance among loggers and woodlot owners and has received the broadest certification given by the USDA Forest Service.[40] Instead of simply cutting from the back toward the notch, as OSHA recommends, bore cuts would be made from one side of the trunk starting in the middle and proceeding away from the notch, as illustrated in Figure 30. The bore cut stops short of cutting all the way out of the side of the trunk opposite the notch. By stopping short, the bore cut creates another hinge that gives the logger more control over the fall, especially useful on trees already leaning in that direction.

Leaning trees may start to fall before the back cut recommended by OSHA is completed; that is, the tree can end up falling faster than the logger can cut from the back toward the notch. When this happens, the tree will split along its length as it falls: this is called the Barber Chair effect, shown in Figure 33. Splitting damages the trunk and reduces the amount of lumber that can be sawn from the logs. With the bore cut, the tree will not begin falling until the final release cut is made at the back. This method gives the logger complete command over the timing of the fall. In addition, by preventing the trunk from splitting, bore cutting makes for a more predictable and there-

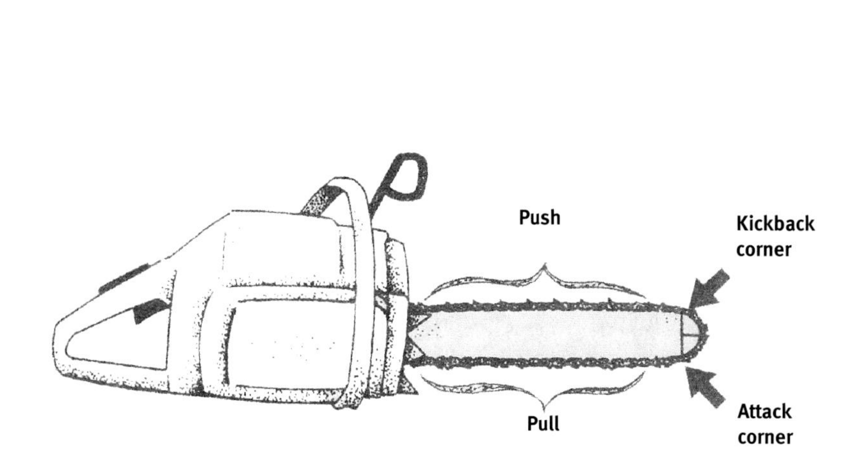

FIGURE 31.
The chain rotates clockwise on the bar. The teeth on the chain cut away from the body of the saw on top of the bar and toward the body on the bottom. Push cuts start beneath a log or limb and proceed up to the top whereas pull cuts start at the top and proceed down. The saw must be pushed away when cutting from beneath and pulled when cutting from the top. Kickback is caused by an attempt to cut from the top part of the tip of the bar—the kickback corner shown above. Cutting from the tip should start from the bottom tip of the bar—the attack corner shown above.
Arborists' Chain Saw Safety Training Guide

fore safer fall as well as preserving the lumber value of the wood.

Bore cutting requires considerably more skill than just a straight cutting from the back and FISTA's position is that it should be used only by those trained to use it. The bore cut starts from the bottom or attack corner of the tip of the saw bar (shown below in Figure 31). Once both sides of the bar are surrounded by wood, the saw is straightened and the bore cut continues: all the way through if the bar is longer than the diameter of the trunk, or a bit more than halfway through if the length of the bar is less than the diameter of the trunk. The bore cut is then made toward the back of the trunk, away from the notch but not all the way out the backside. When the bore is complete, a hinge will be left between the notch and the front of the bore. The thickness of this hinge should be approximately equal to ten percent of the diameter of the trunk. Stopping short with the bore cut leaves another hinge at the back of the trunk as well.

The release cut, the one that allows the

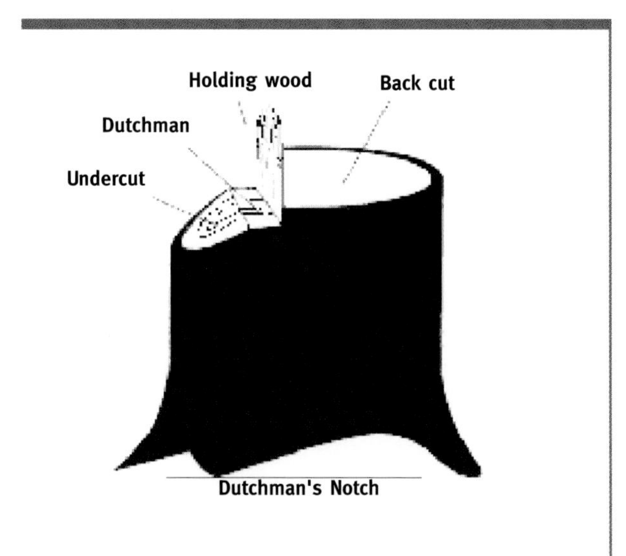

FIGURE 32
A dutchman can cause kickback and splitting along the trunk as the tree falls.
OSHA Logging Advisor

FIGURE 33
The fall of the tree is stalled on the dutchman.
OSHA Logging Advisor

tree to fall, is then made from the back just under the bore cut. As the release cut proceeds beneath the bore cut, what little wood there is between the two gives way and the tree falls toward the notch. If necessary, the fall can be initiated using wedges inserted into the sides of the bore cut. If you plan to do much of your own felling, this method is well worth learning because it is efficient and it provides an extra margin of safety by providing more control over felling. However, bore cutting is not as easy as it might seem from my brief description: proper training is absolutely essential.

As illustrated in Figures 32 and 33, a flat place called a "dutchman" is created when the ends of the first two cuts do not meet as, for example, when the top cut extends well beyond the bottom one. The dutchman prevents the notch from closing completely and thereby interferes with the way the tree falls. As a result, the tree may split along its length as it falls (the barber chair effect), destroying usable wood in the process. Or, the tree may start to fall and then stop at the point where the trunk comes to rest on the dutchman (Figure 33). This is a very dangerous situation since the tree is balanced precariously on the stump, held in place only by its own weight and, perhaps, a thin hinge of wood. OSHA recommends using a machine (a bulldozer or crane) to safely topple a stalled tree.

Tree service companies often use cranes to fell and remove trees. The steel cable from a crane is attached securely around the trunk near the top of the tree (sometimes the tree is topped first to eliminate the weight of excess foliage and limbs). Then the base is straight cut across the entire trunk. The crane operator applies just enough pull on the tree to keep it from pinching the chainsaw blade. Once cut, the entire tree can simply be lifted up and over to a landing zone or into the bed of a truck where is it bucked and limbed.

In wooded areas where there are trees of different ages and sizes, the larger tree being felled might land close enough to the smaller ones to bend them over close to the ground. Trees caught like this are referred to as spring poles. If not cut properly, they can suddenly and very quickly spring back, injuring anyone standing nearby. To solve this problem, the wood under com-

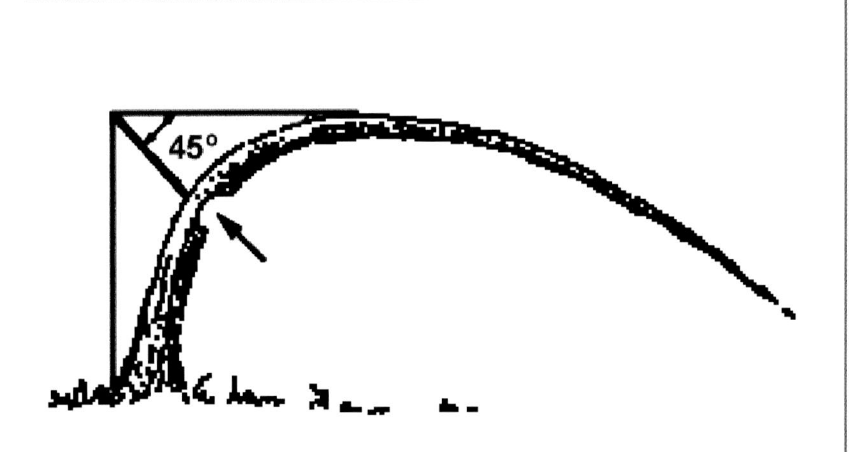

FIGURE 34. Sighting a perpendicular line from the base of the tree up to about where it intersects a horizontal line from the highest point on the trunk forms a right angle. Bisecting the angle (at forty-five degrees) indicates approximately where wood should be removed on the compressed side of the trunk.
OSHA Logging Advisor

FIGURE 35
Cherry tree being topped by Madison Tree Company personnel.

FIGURE 36
Trunk being being lifted by telescoping crane.

FIGURE 37
Cut at base.
Photos by Sam Sherrill

pression should be carefully shaved away where the arrow points in Figure 34 below. This should relieve enough tension on the opposite side to allow the tree to bend slowly or crack without recoiling.

If several trees are being felled and one happens to land on part of the trunk of another, the trunk of the second tree could be under pressure from the weight of the first one. This is known as top lock. Making a first cut all the way through the tree under pressure could be dangerous because one or both sections of the trunk could kick back, possibly striking you with a deadly force. OSHA recommends the trunk be first partially crosscut where it is under compression. The second partial crosscut is then made on the opposite side of the trunk where it is under tension. The two cuts should be made past one another and at least three inches apart. Knowing exactly when and how to make this cut is where professional experience counts. I would be inclined to call for professional advice under these circumstances.

Tree Limbing and Bucking

Limbing

Once the tree has been felled, the next step is to limb and buck it, that is, cut the limbs off and cut the trunk into saw-log lengths. With considerable care, you can do these tasks yourself.

However, you might not have to if the company taking the tree down has the necessary heavy equipment to do the job for you or is going to buck as part of their contract with the property owner. As shown in Figures 35, 36, and 37, by the time the cherry tree was felled it had already been-limbed. All that remained was to buck it into saw-log lengths. When this same company cut a hard maple tree down on my property, they used a crane to lower it onto the bed of a truck where it then could have been cut into saw-log lengths. Once cut, the logs are ready to be hauled to the mill. In my experience, this is the safest and least expensive method for preparing the trees for the mill. The extra time needed to buck the tree into saw-log lengths should roughly offset the time needed to cut the tree into smaller firewood lengths. You may have to pay the company to drop the logs at the mill before going on to the landfill with the rest of the green waste. Smaller tree companies without such equipment cannot lift, cut and haul saw logs. They will want to cut the logs into small sections that are easier to manually load into their trucks. For this reason, you are better off dealing with well-equipped companies.

While limbing and bucking are not as dangerous as felling, there are still significant risks to both tasks that you must keep in mind if you decide to do this yourself. In what follows, I describe the basics of cutting tree trunks into good quality saw logs and the safety procedures you must follow. Even if you do not do the limbing and bucking, you will want to discuss, with those doing the work, how the tree should be cut, both to maximize its value as a saw log and to expedite sawing.

Small and readily accessible limbs should be cut close to the trunk, as illustrated in Figure 38. Rolling the logs and running them through the mill will be much easier if there are no protruding limb stubs. Manually rolling a log with stubs onto a mill is about as easy as rolling a square wheel uphill. Once on the mill, positioning and turning such a log is difficult and time-consuming because the stubs often snag on the mill bed, especially when trying to position the log lengthwise on the bed.

Cutting limbs trapped beneath the tree is more difficult. There are two hazards I have encountered. First of all, from the weight of the tree, the trapped limb is under tension or compression, or both. Once cut, some part of the trapped limb may suddenly spring loose and strike you with surprising force, often in the face. Being hit in the face, especially in the eyes, could be serious. Recall a time when you were hiking through the underbrush and the person in front of you bent and then released a small branch that flew back and hit you in the face. Now think of a larger limb hitting you with even greater force. At the very minimum, safety glasses are a must. As discussed below on safe chainsaw use, the safest and most convenient choice is a helmet with an attached faceplate that covers your entire face, not just your eyes.

When you cut limbs beneath the tree there is also the possibility that the tree itself will suddenly drop and roll toward you. Even if the tree rolls away, your saw bar might be pinched by the limb and the saw might be jerked out of your hands, or worse, pushed toward you. In assessing what cut to make and where to make it, one option for a tree resting above the ground on one or more of its limbs is to cut the trunk (starting with an underside cut) before the limbs. Once the downward pres-

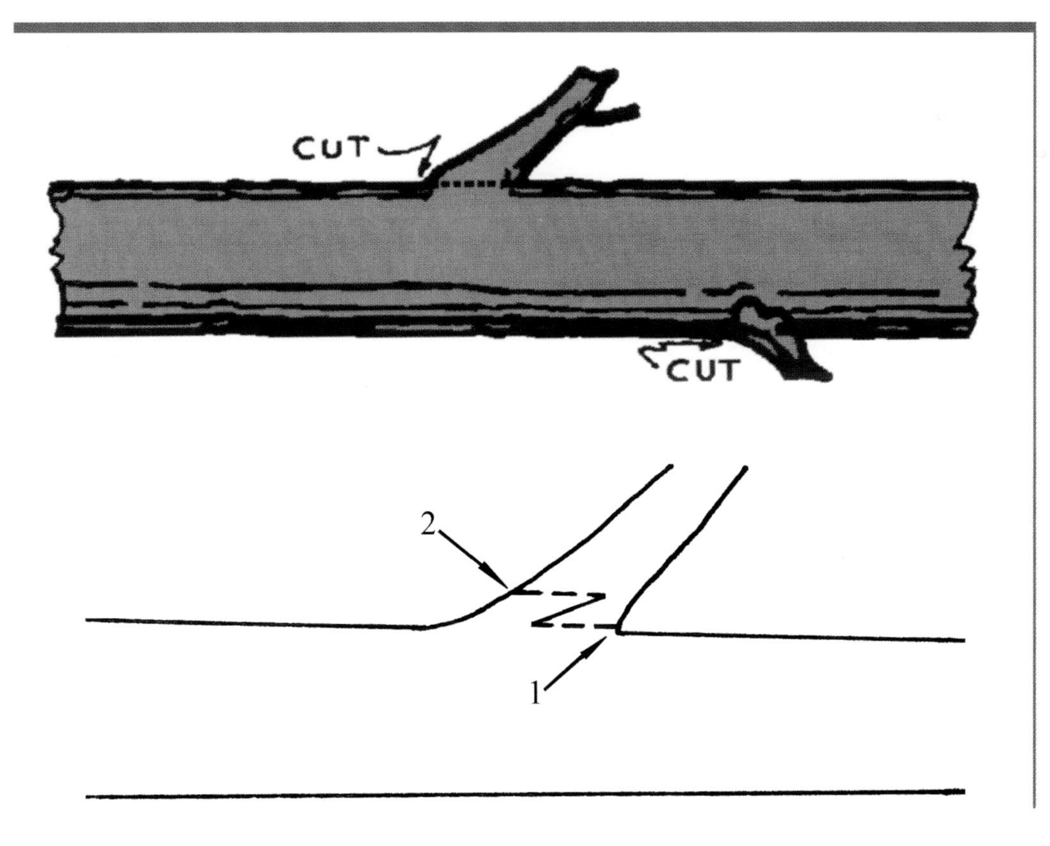

FIGURE 38. Cut limbs close to log.
From Recycling Municipal Trees

FIGURE 39 Limb lock cuts relieve pressures on limbs that would otherwise spring back and possibly injure the chain saw operator.
OSHA Logging Advisor

sure is off the limbs, they can be safely cut.

A tree can fall in a way that presses one or more of its limbs either toward, away from, or sideways to the trunk. If the limbs are cut without regard to the particular pressure they are under, they can suddenly and forcefully spring back. Anyone caught behind them could be injured if struck. At the very least, the chain could be pinched and the saw bar caught in the cut.

To safely release the pressure, OSHA recommends a limb lock cut. The first cut is made where the limb is under compression pressure. As shown in Figure 39, if the limb is being pressed toward the trunk then the first cut, as indicated by the broken line, comes at point 1 and the second cut at point 2 where the limb is under tension.

If the limb is being pulled away from the trunk then the sequence of cuts is reversed: the first cut is made at point 2 and the second at point 1.

If the limb is being pressed sideways, the first cut is still made on the compressed side followed by a second cut on the tensioned side. In all three cases, the cuts are made past one another but at different levels.

Often, when land is being cleared for construction, whole trees are pushed into unorganized piles by heavy equipment. **Cutting in a pile of interlocking limbs and trunks is very hazardous.** Having done it once myself, I strongly recommend not doing it. If you are cutting from the top, the logs may shift beneath you once you have made a cut. You could easily fall while holding a running chain saw. If you are cutting down in the pile, you could be pinned by the sudden shift of a log or a limb. You are likely to be severely injured if struck by or caught between two logs weighing several thousand pounds apiece.

The safest procedure is to pay the heavy equipment operator to push the pile apart so that cutting can be more safely done. I paid the loader operator to disentangle a pile of trees so I could cut them. While he was working, a log about eight feet long and six inches in diameter flipped out of the pile, over the loader bucket, and flew straight through the windshield, coming to rest a short distance from the operator's head. Even heavy equipment does not provide complete protection from accidents of this sort. Better to arrange the trees in parallel rows or stacks so that they do not become entangled with one another to begin with.

Bucking

After it has been felled and limbed, a tree trunk is then bucked. Bucking means cutting the trunk either into commercial saw log lengths or into nonstandard lengths for special projects.

While, ordinarily, bucking refers to cutting a tree lying on the ground, I'm going to extend the meaning of the term to include cutting the trunk into log lengths after it has been limbed but before it has been felled. Urban trees often grow in very confined spaces bordered by sidewalks, streets and buildings. To protect people and property, the tops and limbs are removed and then the trunk is cut and lowered by crane or dropped in sections. Unlike felling trees by hand in a forest or in an open pasture, tree service companies often cannot fell tall street trees in one piece.

Saw log lengths are determined by the overall length of the trunk, curvature or sweep in the trunk, commercial lumber standards, specific project requirements, or some combination of these factors. Except for a few highly specialized mills that can cut logs with some sweep, sawmills only cut in a straight line and veneer mills only peel straight logs. Therefore, logs must be as straight as possible to maximize the amount of lumber and veneer that can be recovered.

An example of the importance of sweep is shown in Figure 40. After the ash tree in the picture is topped where the branches fork, the trunk should be felled and then bucked as shown in the drawing on the right. A lot of lumber is lost when trees such this one are bucked without regard to sweep. Of course, all potential lumber is lost when the trunk is bucked into short sections (2 to 3 feet) for easy loading and hauling. This is typically what happens to trees when they are seen as green waste and not as valuable saw logs.

Standard lengths for commercial cutting start at 8 feet for both softwoods and hardwoods. Lengths for softwoods progress in 2-foot increments: that is, from 8 to 10 and on through 16 feet. Hardwoods are cut both in 2-foot increments and in uneven lengths. The log lengths are actually at least 2 inches longer than the standard. The extra several inches, the trim allowance, compensates for splits at the ends of the log and for the occasional uneven end cuts. In general, sawmills end trim lumber to the standard lengths. I tend to leave 3 to 4 inches to compensate for

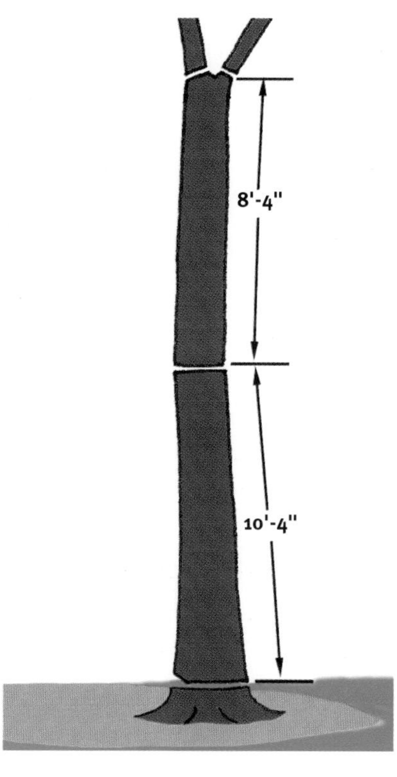

FIGURE 40 and 41 The cuts in the diagram on the right indicate how the ash tree pictured on the left would be bucked into two logs, ten and eight feet in length (plus a four-inch trim allowance), to obtain the maximum amount of lumber. **Photo and drawing by Sam Sherrill**

splitting during drying, and for final trimming to length after I have run the boards through my thickness planer. The planer occasionally snipes the leading end of the board (that is, the planer cuts slightly deeper as the board starts through). Having left a few extra inches allows me to cut out the sniped part of the boards.

You can, of course, cut lengths shorter than 8 feet as well. However, most sawmills will not accommodate logs shorter than 3 to 4 feet.

As long as the diameter of the log is not too great, shorter-than-minimum segments can be cut on a shop band saw. I used my band saw to cut small pieces for box lid panels from a short piece of curly walnut I found in a bundle of firewood. For repeated resawing, a sled can be easily made that slides in the miter groove on the band saw table.[41] Short logs are clamped into the sled for safer and more convenient cutting. If the diameter of the short log down the middle is greater than the maximum cutting capacity of the band saw, you can trim it on all sides until it fits. By repeatedly turning and trimming, you eventually end up with a mini-cant (a cant is a log that has been squared up on at least two sides).

Outside of commercial cutting, length depends on what you have in mind for

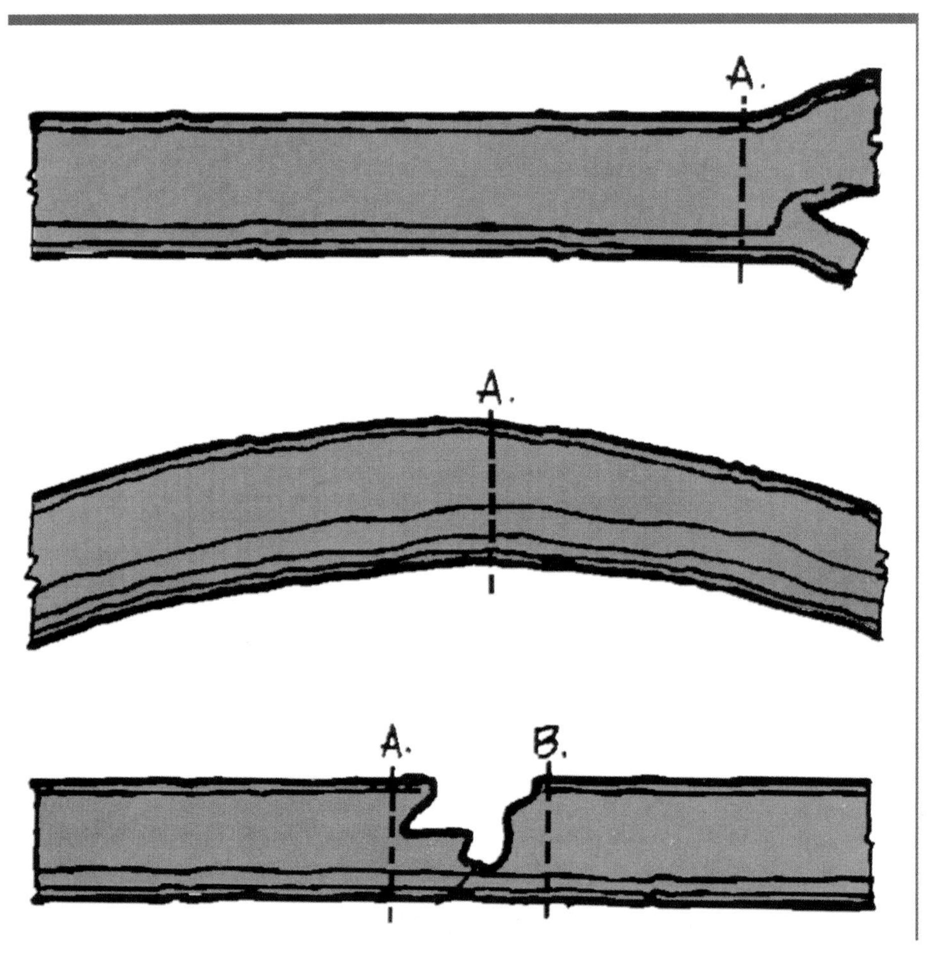

FIGURE 42
To meet commercial standards, the log would be cut at A. By artistic standards, the crotch to the right of A might be left on the log so that the crotch wood is part of the resulting lumber.
From Recycling Municipal Trees

FIGURE 43
This log would be cut at A to minimize sweep.
From Recycling Municipal Trees

FIGURE 44
This log would be cut at A and B to eliminate the defect between them.
From Recycling Municipal Trees

the lumber ultimately cut from the log. For most of the logs I cut, I view bucking as the first stage in the construction of a specific project. For example, if I plan to build a 10-foot dining table then clearly I need to cut the log into one or more 10-foot (plus) lengths. (Using boards cut from the same log gives me the option of using matching grain patterns for the tabletop.) In this example, the project defines the lengths cut.

However, I also look at the log to see if there is anything about it that suggests a project that I do not have in mind. That is, instead of cutting the log to fit a predetermined project, I look for the log to suggest a project, and then cut with that in mind. For example, in Figure 42, the cut would be made at point A to maximize the commercial quality of the log. However, as long as the crotch is not badly split or rotted, it could be left on. At one end, the resulting lumber would have unusual grain that only appears in crotch wood. While leaving the crotch on the log decreases its commercial market value, it could increase its artistic value. The boards that go into a unique piece of furniture ultimately may be worth more than they would have been worth had they been cut the usual way.

I also look for burls on tree trunks and roots. Burls are large bulges or semi-hemispheric growths that protrude from the side of the tree trunk or from the roots.[42] They often grow on trees that are stressed. When they are not decayed, burls are highly prized for their intricate grain. Burl wood often shows up on the dashboards of luxury cars, as gunstocks, or as expensive turned bowls. There is a substantial market for them, especially in walnut, among woodworkers and veneer manufacturers.[43] Since they are very valuable, I would cut to preserve them whole even if I have to sacrifice length in the remaining logs. For the past fifteen years, I have been saving a solid walnut burl cut from a yard tree in southern Indiana. I'm still looking for a project worthy of it.

As illustrated in Figure 43, cutting at A can reduce the curvature, referred to as sweep,in a tree trunk. Keep in mind that even these shorter logs may contain stress wood that will split, cup, or warp during drying.

To increase the commercial value of logs, tree trunks must be cut to minimize defects such as limbs, rot, and cavities. As illustrated in Figure 44, cuts should be made at A and B to eliminate the gaping hole in between. However, good lumber can still be cut from logs with slight defects near the bark. I would sacrifice a few boards to get greater length in the remaining ones. Also, not all defects are defects if you look at them as opportunities to create something different. Splits can be tied together with a butterfly key and small pockets of rot can be filled in with tinted epoxy. Again, what may be commercially unacceptable could be artistically desirable.

There are several basic rules to follow when bucking a tree into logs. First, you must follow the rules for safe operation of the chain saw as discussed in the next section.

If the log is lying flat on level ground, then you can cut it from the top down, stopping just short of cutting into the ground itself. Soil, stones, and other debris will quickly dull the cutting tips of the chain, so you want to avoid running the saw into the ground. This cut leaves a slight hinge of wood. I use a peavy or cant hook (see Figure 45) to break the logs loose from one another or rotate them just enough to finish the cut without touching the ground. If the log is not too heavy, I use a log jack to raise it up off the ground (this is a handy way to cut firewood as well). The log carrier and the lifting tongs can also be used to lift logs into place for easier and safer cutting.

If the log is not lying on level ground, I wedge a rock or a chunk of wood between the log and the ground to prevent the logs from rolling toward me after I finish the cut.

A felled tree may end up straddling a swale or depression in the ground or lying partially on a hillside. In this case, because it bridges an open space, the trunk is compressed on the topside and is under tension on the underside. A single first cut downward from the top will result in a pinched bar and chain when the two halves are pressed together by the weight of the trunk. To avoid pinching, you can first make cuts on both sides, then the top, and finally up from the bottom. Cutting a wedge from the top instead of making a single cut should further reduce binding.

If just one end of the trunk is suspended without any support, then compression and tension are reversed: there is compression on the underside and tension on the topside. In this case, you make side cuts, then make a single cut, or cut a wedge on the underside of the trunk and the final cut from the top.

In both cases, be especially mindful of how the logs might drop and roll once freed by the cut. You should not cut with the saw directly in front of you but should cut to one side of your body. If the log drops and rolls toward you, and you are cutting directly in front of yourself, then the log could push the saw between your legs. You do not want a saw, with the chain perhaps still running, and the log to be competing for the small space between your legs. It pays to think of these possibilities in advance because, once they start to happen, you will not have time to review the options and then choose the safest response.

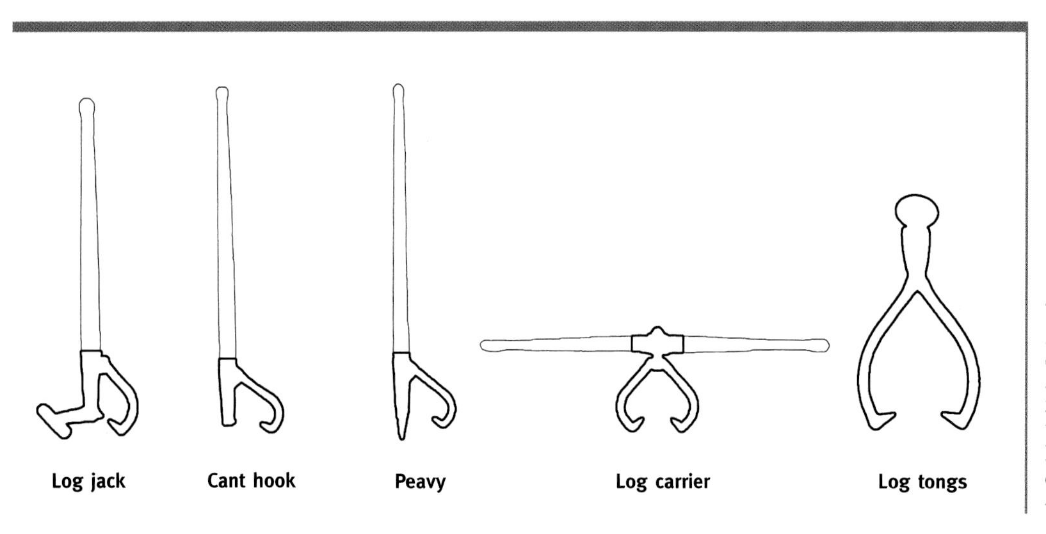

FIGURE 45. From left to right: log jack, cant hook (no point at metal end of handle), peavy (a cant hook with a point), log carrier, and log tongs.

Another point to keep in mind when limbing and bucking on private property is to do as little damage as possible to the owner's turf, other trees and plants, or structures. You should clean up all debris, including sawdust. In nearly all cases, property owners, especially those who have invited you in to retrieve the logs, will be accommodating about any minor mess created by cutting. However, you might want to ask an attorney to draw up a release form for the owner to sign in advance. Also, photograph the site before you start and then again right before you leave. The photos can serve as documentation of what you found when you arrived and the condition of the site when you left. If there is a dispute later on about damage or cleanup, you will have visual proof of what you did, and did not do, as well as a signed release.

One other point to keep in mind when you are invited onto someone's property to cut: make sure both you and the owner are very clear about property boundaries. Several years ago, a property owner in my area wanted several cherry trees felled that were growing in what supposedly was his backyard. He was not present the day the trees were felled. Unfortunately, one member of the group I was with that day insisted on felling a large cherry that turned out to be on an adjacent piece of property (this is also the tree I mentioned above that fell the wrong way). The owner of the adjacent property was very upset when he discovered that one of his largest and otherwise healthy cherry trees had been felled and cut into lumber. Initially, he threatened to take legal action, including filing charges of trespassing and malicious destruction of property. Ultimately, he was mollified after I gave him all of the lumber that had been cut from his tree plus some extra cut from the other trees that were not his. You do not want to have this experience, so be clear about where you are cutting. Take photographs of boundary markers or stakes, in case there is a dispute later on

Safe Chain Saw Operation

When using a chain saw, you should keep foremost in your mind that it is considered the most dangerous of all the hand-operated power tools. For the year 2000, the U.S. Consumer Product Safety Commission estimates that there were over 26,000 chainsaw-related injuries across the country.[44] A powerful saw with a sharp chain can quickly and easily cut through logs and limbs; that is, after all, what we want it to do. Even the hardest of woods yields to the chain saw, your flesh and bones are no challenge at all.

Personal Protective Equipment

Safe use begins with wearing the right protective equipment. Starting from the top, an ANSI-rated hard hat with attached protective face screen and earmuffs is essential, and it is convenient. The hard hat protects you from falling branches and other debris from above while the earmuffs protect your hearing from the sound of the chain saw. I use both earmuffs and flexible ear plugs because I particularly dislike the sound of the saw. (In operation, a chain saw can reach 115 decibels, well above the 85 decibels that from continuous exposure will produce permanent hearing loss.) The face screen not only protects your eyes and face from injury but also protects glasses from damage. The lenses in my glasses are plastic and are easily scratched by flying wood chips. I wear cut resistant gloves and leg chaps (which are now being made with layers of Kevlar, the material used in bulletproof vests). The chaps should prevent serious injury if the still turning chain should come in contact with the front of my legs or my ankles. I also wear waterproof steel-toed shoes that cover my ankles both for protection from the saw and from logs and limbs that might fall on my feet. Neither ordinary leather shoes nor running shoes offer anything close to the same protection as steel-toed shoes.

Wearing all this gear at once, especially on a hot humid day, might not seem very comfortable (though many of the cut-resistant chaps and pants are very comfortable). However, whatever discomfort there is pales by comparison to how you would feel if you cut yourself with the chain saw. Complain about it but then wear it.

Finally, have an OSHA-rated first aid kit and a cell phone on-site and readily available. You should have someone present with you to administer first aid for minor injuries and to call the local EMS in case of a more serious injury. Like swimming alone, using a chain saw alone is risky behavior.

OSHA LOGGING ADVISOR

Chain Saw Injury Locations

Notice how most injuries occur on the lower left leg and the left arm. Be sure to protect those areas well.

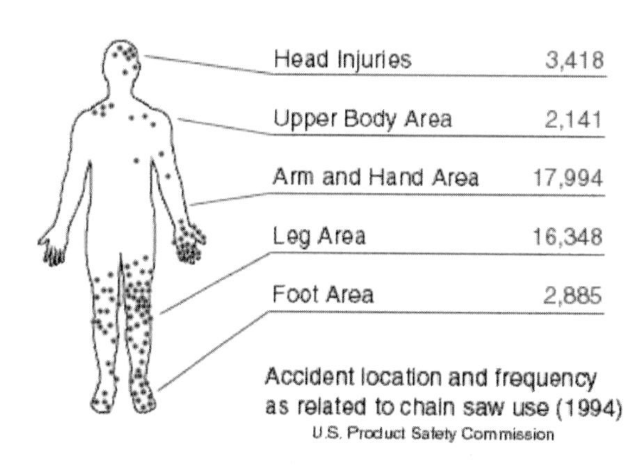

FIGURE 46
This chart graphically illustrates where most chainsaw injuries occur. Notice that most of the injuries cluster on the left hand and on the legs, somewhat more on the left leg than on the right one.
OSHA Logging Advisor

Chainsaw Safety

The chainsaw owner's manual describes how the saw should be safely started, used and maintained. At a minimum, you should know how to keep the saw in good working order to minimize wear and tear on yourself and on the machine.

There seems to be no shortage of advice in magazines and on web sites on how to start and cut with a chain saw. I have observed that employees of tree service companies have their own methods as well. Rather than trying to weave safe instructions from all these various sources, I am again recommending standards set by OSHA and methods taught and used by FISTA (the Forest Industry Safety and Training Alliance).[45] Even with these instructions, you should still carefully read the manual and follow its instructions on fueling, maintenance and safe operation of your own machine.

Before starting the saw, make sure it is in good running order. The chain should be adequately tensioned and all the parts should be securely tightened. A loose bolt flung by a turning chain becomes shrapnel that can tear skin or injure the unprotected eye of an observer.

Clearly, you do not want to be smoking while fueling the machine nor do you want to be near a fire—one started to keep you warm, for example, in cold weather. You should be no less than twenty feet from open flames and should start the saw at least ten feet from the place where you put gasoline in it. This rule may seem overly cautious but, in addition to being burned, perhaps severely, think how stupid you would feel if you caught fire even before using the saw.

There are two ways recommended by FISTA to start a chain saw. For both, the saw chain brake is set first, then the ignition switch is set to the on position and the choke is set for a cold start.

The first starting method involves firmly holding the saw on the ground with your foot on the back handle and your arm stiffly holding the front handle bar. Then the starting rope is pulled. In my experience, the only drawback to this method is that occasionally, when I pull the rope, the front of the saw tips forward enough to strike the ground or pavement. I place the saw on a couple of wide boards so that if the front does tip forward, the chain hits wood and not dirt or pavement.

The second method entails firmly clamping the rear handle of the saw between the upper part of your legs just above and behind your right knee. Grip the front handle bar with your left hand on the upper left corner at a point in line with the starter rope. Make sure there is adequate clearance between your leg protectors and the chain. Use your right hand to pull the starter rope in a short, quick action.

I generally use the first method when the saw has not been run for some time and is, for me, more difficult to start. I use the second method to restart the saw after the engine has warmed up.

There are two other commonly-used ways that are not recommended by FISTA. The first is the drop or air start: holding the saw by the handle up close to the body with one hand while the gripping the starting rope with the other hand. The saw is started by quickly pushing the saw away and down from the body with one hand while simultaneously pulling on the rope with the other. The other method involves placing the saw on one knee and pulling the rope. I do not recommend either of these methods because both could inflict severe cuts on your leg. Look at the number of leg injuries in Figure 46.

Do not touch the blade when the sawing is running. While this seems as obvious as being told to remove the wrapper before eating the candy bar inside, look once again at Figure 46. At least some of those left-hand injuries had to have been caused by the operator thoughtlessly or inadvertently touching the moving chain. The larger point, of course, is to remind you to keep your hands away from the bar and chain. Unless the motor is off and you are tensioning or changing the chain, there is no reason for your hands to be anywhere near the front of the saw, especially if it is running. You are endangering your hand if you reach out to the side or front of the saw, while it is running, to clear debris or push a cut or dangling limb aside.

By OSHA standards, you should not carry a running saw more than 50 feet or over hazardous ground (for example, steep hillsides) without applying the chain brake. The safest procedure is to either shut the saw off or apply the brake when you are not immediately using it to cut.

Unlike a circular saw, a chain saw cannot be operated safely with one hand. By its weight alone, few if any of us using just one arm and hand could hold a running chain saw steady and parallel to the ground. So, use both hands and a firm grip on the handlebar to hold the saw while cutting. Stagger your feet and bend your knees slightly so that you are in a flexible position in case of kickback.

Do not overreach to cut, especially from a ladder. Do not cut above shoulder height. If you should happen to fall over backwards, the saw will fall toward your body, very possibly into your face. Do not cut with the saw directly in front of you in case the chain or bar binds, and the saw either kicks or pulls back toward you.

I try to stay focused on the cutting task at hand. If my attention flags from fatigue, or more likely, from sweat and sawdust on my glasses, I try to take that as the moment to stop and rest. Rushing to finish near the end of the day or stretching your luck by taking a chance you know is risky "just this once" will probably backfire on you. In general, if you feel uncomfortable about what you are doing, or about to do, then trust your own instincts and do not do it.

Finally, pay close attention to the feeling in your hands, wrists, and arms. Prolonged exposure to the mechanical vibration produced by a chain saw can damage the blood vessels in your hands and fingers. In turn, this can lead to skin, nerve, and muscle damage. The condition is called hand-arm vibration syndrome (HAVS). It is also referred to as white finger, dead finger, or Raynaud's Syndrome. Symptoms are a tingling sensation or numbness in the fingers, white or blue fingertips, difficulty picking up small objects, trouble with buttons and zippers, or a diminished ability in the hands to feel heat, cold and pain. Caught early enough, the damage is reversible. Prolonged exposure can lead to permanent physical impairment and disability. Cold temperatures and smoking both exacerbate the condition.[46] To reduce the risk of acquiring this problem, do the following:

1. Take a ten-minute break for every hour of saw use.
2. Keep the saw in peak working order to minimize vibration. If you are in the market for a new saw, look for one that has a built-in anti-vibration system, is relatively lightweight, and has a high power-to-weight ratio.
3. Try to minimize the amount of actual sawing time, thus your exposure to vibration. Alternate-hour sharing of the cutting load will help.
4. Do not operate the saw at constant full throttle.
5. Let the weight of the saw do most of the work, as opposed applying your body weight to it. Pressing or pushing on the saw increases the pressure on your wrists and hands.
6. Use as loose a grip as you can while still safely controlling the saw. A tight grip both restricts blood flow and allows more of the vibration to be transmitted to your fingers and hands.
7. Stop using the saw if you notice any of the symptoms listed above.

Carpal tunnel syndrome (CTS), damage to the median nerve in the hand, also causes pain and loss of feeling, especially in the thumb and first three fingers. This is a rather common repetitive motion injury that arises from continuous and repetitive use of the hands and wrists, as happens with the extended use of a chain saw to buck and limb trees. Caught early on, the damage is reversible. Untreated, CTS leads to permanent nerve damage. Like HAVS, if you feel pain and lose feeling in your thumb and fingers after using a chain saw, stop using the saw. In either case, if the symptoms persist, you should see a physician.

Summary

To turn trees into logs in urban areas, we need to know how many trees there are, and how many are being removed. Until Forest Service urban forest surveys begin providing us with data that cover many years, we will have to rely on local and state sources for information about the number and kinds of trees there are in our own cities and towns.

Trees and lumber are broadly divided into hardwoods and softwoods, a classification that has less to do with the actual hardness of the lumber than with the ways these two types of trees reproduce. Trees grow in every direction. That they grow outward means that they grow over and eventually around anything they encounter, from wire fencing and spikes and bolts to concrete and even spoons and beer cans. This is important to know when felling and especially when sawing logs into lumber. While metal detectors can detect metal, they cannot detect glass, concrete, stones and other nonmetallic objects.

There are ways to minimize accidentally cutting into embedded objects; for example, avoiding double-stemmed trees that are likely to have been cabled together is one way. Also, band saw mills that use inexpensive blades keep blade repair, replacement and mill downtime costs to a minimum.

A distinction is drawn between the sapwood and heartwood of most trees. Generally, though not always, heartwood is darker in color than sapwood. The accumulation of extractives such as tannin accounts for the darker color. The difference between sapwood and heartwood is pronounced in walnut and cherry but not discernable in buckeyes and most spruces. Usually the heartwood is used for lumber because of its color, strength, and resistance to decay.

Adverse growing conditions, storm damage, insects, disease, and the way trees are knocked over by heavy equipment can all reduce the amount of lumber that ultimately can be cut from the logs. Limbs that grow parallel (or nearly so) to the ground, as well as trees that did not grow perpendicular to the ground, are likely to produce reaction wood. The resulting lumber tends to severely distort during drying and, once dry, might not take finishes well. Storm damage, insects, and disease tend to destroy otherwise usable heartwood. Even though usable wood may be present, disease-infested trees may be destroyed to halt further contamination.

Spalting produced by fungi, if caught before the wood is damaged, yields lumber that is highly prized for boxes, door panels, and turned bowls (do consider the possibility that you could have an allergic reaction to the fungi that cause spalting).

Damage from felling by heavy equipment can be minimized by partially uprooting the tree first and then pushing it over instead of just pushing on the trunk until the tree is ripped out of the ground.

Tree felling, limbing and bucking are very hazardous. You can be injured by falling trees or limbs, and by the chain saw itself. I strongly suggest you leave felling, especially in crowded urban areas, to professionals who have the skill, the equipment, and the necessary insurance. Amateurs can do bucking and limbing as long as strict precautions are taken with the location where the logs and limbs are cut, and the way the chain saw is fueled and used. An even better arrangement is to have the professionals, who are removing a tree, buck it into saw logs as they take it down.

You must wear the proper safety gear and use the chain saw in a safe manner to avoid serious injuries.

As I pointed out in the Introduction, making the best use of fallen urban trees is worthwhile as long as you do not injure yourself in the process.

3 FROM LOGS TO LUMBER

Introduction

In the latter part of the previous chapter, I described the basic procedures for safely felling, limbing and bucking urban trees. Once that is finished, the logs are ready for the sawmill.

The next step is to transport the logs to the mill, or the mill to the logs. Then the logs are cut into lumber and the lumber is dried. Transporting, cutting costs and cutting options, and drying, from the viewpoint of the log owner, are the subjects of this chapter.

In most urban areas, professional crews working with the right equipment will remove trees from parks, streets, utility company and other commercial properties. Arrangements can be made to have the logs hauled to the sawmill or to have a mill set up on-site when there is time, and space in which to cut. I focus here on moving logs on private noncommercial property where access is often difficult, and equipment options for small jobs are limited. These logs most often come from residential trees that grow in the yards of homeowners.

Next, I review the basics of sawing logs into lumber and, finally, drying the lumber to the appropriate moisture content (hereafter designated as MC).

Skidding and Yarding

Whether you saw them into lumber on-site or off, you will probably have to move the logs from the place where the trees were felled to the mill or to a waiting truck (a landing, in logging terms). In commercial logging, this task is referred to as yarding, which basically means gathering all the logs at a landing for subsequent hauling. Skidding is the term used to describe how the logs are moved from their stumps to a landing. They are either carried or dragged across the ground from the cutting to the loading sites (look again at Figure 26 in the last chapter to see a skidder used in modern commercial logging).

Before diesel and steam driven machines were available, teams of oxen, called bull teams, were used to skid logs (the man who guided the team was called a "bull whacker" because he used a long leather whip to keep the oxen moving). A skid road was made by imbedding small logs in the ground at intervals across the logging trail. Tied one behind the other, up to a half dozen saw logs could be pulled down the skid road by the bull teams. The first log would be tapered at one end so that it would not snag on the road logs. Lard or grease was smeared on the skid road so that the logs would slide easily; this was known as "greasing the skids". You will not have access to a bull team, nor will property owners allow you to build skid roads across their lawns (so there is no reason for you to ask others to call you the bull whacker—you can imagine what they might call you instead). This leaves you with two basic options.[1]

First, see if the tree service company will move the logs, either manually or with a crane or knuckle boom. Unless the logs are being left on-site for the owner to dispose of, the tree company has to move the logs anyway as part of the job. If the logs must be hauled to the mill, then for an extra payment the company might be willing to cut them to saw log lengths, load and then haul them. This is the ideal solution for small projects where you are working on your own. I especially like those jobs where the truck can be parked next to the tree being felled.

You will have to do the skidding yourself if there is no tree service company involved and the tree is not near a street or driveway. Trees downed by storms will have to be limbed and bucked before being moved. Even trees left by tree services may have to be bucked into manageable saw logs. In tight urban spaces, the size that can be rolled or lifted is an important determinant of log lengths.

Many years ago, when bull teams were used to move logs, skidding was done with little concern for the soil or vegetation around the road. Skidding logs across someone's lawn today is an altogether different matter. Unless carefully done, logs rolled or dragged across a yard can damage grass, garden plants, and the property owner's disposition. The following are some suggestions on moving the logs without hurting yourself or damaging property.

Before you start, take photographs of the area where you will be working. Being seen taking the photographs demonstrates to the property owner that you are thinking ahead about questions that could arise about damage. Also, you might ask an attorney to prepare a release that covers minor yard damage. Photographs and a signed release should afford you some protection in case questions arise later on about damage.

Protecting Yourself

You will need the help of several others. Moving logs alone is like swimming alone: if you get into trouble, there is no one to help or to go for help. You might be able to roll logs on your own but there is no way you can lift them by yourself. When lifting, consider this: a freshly cut 6-foot red oak log 24 inches in diameter weighs approximately 1,200 pounds. At 18 inches in diameter, the log weighs about 650 pounds. Even a log 12 inches in diameter weighs about 300 pounds. You can calculate the weight of a log by multiplying its length times the weight of a one-foot section given in the table in Appendix B. Weights are by species and diameter (a one-foot section of red oak 24 inches in diameter weighs 198 pounds; multiplied by 6 equals the near 1,200-pound log weight).

Wear thick gloves to protect your hands against the bark and any unexpected encounter with jagged objects embedded in the wood. Be aware that gloves will not protect your fingers from being mashed or cut by the chain saw. However, you can protect your feet by wearing the same above-the-ankle steel-toed shoes you wear when limbing and bucking.

There is no scientific evidence that elastic waist belts will protect your back from lifting-related injuries.[2] Like wearing a string tied around your finger, wear a belt around your waist and lower back if it helps you remember to do the following:

1. Lift logs and lumber with your legs from a squatting position, keeping your back straight and holding the load close to your body. Do not lift with your arms while bending over, or reach and lift while your back is bent. Do not lift into a position that twists your back.
2. Do not lift too much at once. Back and shoulder injuries are common when moving logs. I know from my own experience that recovery from a torn rotator cuff in the shoulder can require months of painful physical therapy. You can also compress one or more vertebrae in your back, leading to painfully pinched nerves.
3. Take rest breaks. Do not lift continuously even when each load seems relatively light. The total load lifted in a day's work can add up to strained, or even injured, muscles.

Illustrated lifting rules are provided in Appendix C. Everyone involved in lifting logs and lumber should first review these rules and the accompanying illustrations.

Minimum Equipment Needed

The following will make skidding easier and safer:

1. Log jack and cant hook (see Figure 45 in the last chapter). My log jack is a two-in-one tool: the T-bar can be removed, converting the jack into a cant hook.
2. Log carrier and lifting tongs.
3. A half dozen short cylinders for rolling logs. I use the cutoffs from lolly columns, steel tubes three to four inches in diameter used in construction to support horizontal floor beams. I retrieve them from residential construction sites.

4. Sheets of plywood, also salvageable from construction sites.
5. A chain and a couple of bricks or small flat stones.
6. A pipe buggy or dolly, available from well-stocked tool rental stores.
7. An electric winch that fits on a bumper hitch and runs off a car or truck battery. I own one because it is so useful. Winches should be available from tool rental stores.
8. A 10-foot by 100-foot roll of plastic sheeting.

Skidding in Urban Areas

Under the best circumstances, you can roll logs sideways in an unobstructed straight line on flat, dry, unfenced ground. When there are just a few logs, they can be rolled to the driveway or street for loading or cutting. Lawn damage should be minimal, especially if the ground is dry. If you have to change directions, the cant hook can be used to push the log onto the brick or stone. Place the brick under the middle of the log so that the log can be easily pivoted to the right direction. Trying to turn a log on the ground by pushing on one end while pulling or holding the other is difficult to do, and can gouge the ground and damage the grass.

Fenced yards (not enclosing driveways) usually have gates about 3 feet wide. Since a log 3 feet in length cannot be cut on most sawmills, a usable log will be longer than the gate opening is wide. One way to get through the opening is to position the log roughly perpendicular to the gate and use the lolly column cutoffs mentioned above as rollers beneath the log. As the log rolls forward, move the column cutoffs from the back to the front. This method can be used to roll logs past obstructions as well. If the ground is wet, the weight of the logs might press the cutoffs into the ground, preventing them from rolling. You can get around this problem by placing sheets of plywood on the ground and rolling the logs and cutoffs over them. This temporary skid road should allow you to move the logs without damaging the lawn.

Small logs can be pulled uphill manually with a hand winch attached to a truck or tree. An electric winch attached to a truck's towing hitch is a much better alternative. A chain, attached at one end to the steel cable from the winch, is fastened at the other end around the front of the log (loggers call this a choker chain). The winch then pulls or skids the log uphill. Use the chain saw to taper the front of the log so it will not snag or dig into the ground as it is being skidded. Instead of tying the chain directly to the log, it can be attached to a pair of tongs. The pressure from pulling will increase the tong's grip on the front of the log. Again, if the ground is wet, a temporary skid road can be fashioned with sheets of plywood.

The cable and chain should run above the ground and straight to the log. You should avoid running the bare cable and chain around buildings, in contact with the ground, or around a tree that is not expendable. The tension created by pulling the log will cause the moving cable to cut into and damage whatever it rubs against. The extra friction can strain the winch and could fray the cable and cause it or the chain to snap. Instead, as illustrated in Figure 47, pulleys can be used as turning points to pull logs around obstacles, a tree in this case. Finally, this is not a job for the family automobile. Trucks and tractors are built for these kinds of tasks.

Before starting, read the safety instructions that come with the winch, chains, pulleys and any other attachments. Also, review the following rules with participants and observers. Once underway, you should closely monitor the process to insure that everyone is observing these rules:

1. Wear leather gloves when handling the cable.
2. Keep your hands away from the point where the cable enters the winch: if caught, your hand could be pulled by the cable onto the rotating winch drum.
3. Keep volunteers and spectators at least a hundred feet away from the cable and log. If the cable breaks, bystanders could be struck by what, in effect, becomes a steel whip, or they could be run over by the rolling log.
4. Never stand next to, or step over, a moving cable or a cable under load: you do not want a chain to break and snap back between your legs.
5. Operate the winch from behind the truck door, another tree, a fence, or

FIGURE 47. An easier way to get a log around a tree and to the top of a hill is to attach a pulley to the tree and use it as pivot point. Once the log is beside the tree, the rope can be attached directly to the winch or, in this case, to the hitch on the truck and pulled the rest of the way.
Illustration used by permission of Sherrill, Inc.

something that will afford protection in case the cable or an attachment breaks.

6. Place a heavy blanket or mat over the cable, about halfway between the winch and the log: this can reduce the whipping action of the cable if it suddenly lets go.
7. Make sure the cable is not frayed and that pulleys, tongs, and other attachment devices are in good shape and securely fastened, and thus will not fail under pressure.
8. Never hook or tie the winch cable to itself, or tie it like rope around the log or the bumper hitch on a truck or tractor. This causes kinks in the steel threads of the cable that could lead to cable failure. Use a log chain or tongs (Figure 45 in the previous chapter).
9. Do not pull a load with the cable fully extended from the winch drum. Follow the specific instructions regarding the number of turns of cable on the drum: three full turns are minimum.
10. Do not exceed the stated load capacity of the winch, cable, chain, pulleys or any other attachments. The lowest load capacity is the weakest link between the winch and the log.
11. Watch the motor to make sure the cable does not pile up on one side of the cable drum or that the motor does not overheat from excess running time. Electric winches cannot be run continuously. During pauses that allow the motor to cool down, check all connections to make sure that the log, cable, chain, attachment points, and pulleys are securely fastened.
12. As an extra precaution, tie one end of a strong rope to the log itself, or to the buggy or dolly, and the other end to a nearby tree or other strong stationary object. This rope will serve as a safety line in case the hauling line or one of its attachments should break.

The pipe buggy shown in Figure 48 can be used to carry a log either across flat ground or, attached to the cable and chain, up a hill. The weight of the log should not exceed the buggy's carrying capacity, and the combined weight of the buggy and log should not exceed the pulling capacity of the winch or the strength of the cable and chain. Be sure to check the owner's guide for your winch to find its maximum dead load and pulling capacity. The buggy shown below will hold up to 1,000 pounds. (Again, refer to Appendix B for log weights by species and diameter.) The log can be loaded by placing the buggy upside-down on the log. Once fastened to the log, the buggy can then be rather easily turned right side up. This eliminates the chore of lifting the log onto the buggy.

A pipe dolly, pictured in Figure 49, can carry heavier logs than the buggy. The one shown below will hold up to two thousand pounds. It is designed to allow one person to carry pipes up to 20 inches in diameter and 20 feet long. Its log-carrying capacity is somewhat less for most species. For example, it will carry a red oak log 20 inches in diameter and 10 feet long weighing about 1,400 pounds. Since the dolly is slightly less than 3 feet wide, it should fit through most backyard gates.[3] I have rented one for about $20 per day, and transported it in my pickup truck (tied down, the two wheels rest against the inside of the tailgate and the long handle rests on a pad on the top of the truck cab).

Logs going to a sawmill will have to be loaded into a truck. Both the buggy and the dolly can be rolled up a ramp into the truck bed where the logs can then be unloaded. Loading this way eliminates the need to manually lift one end of the log onto the edge of the truck bed, then lift the other end and push the log all the way forward. A ramp up to the bed edge can be fashioned from plywood and two-by-fours, or several sheets can be laid on the ground and on the rear bumper (with the tailgate removed). If you do this frequently enough, metal ramps that fit onto the truck bed would be worth renting or purchasing. Several volunteers will be needed to pull the buggy or dolly into the truck bed. One volunteer should be in charge of pulling a rope attached to the buggy or dolly through tie-down rings on the truck bed to keep one or the other from rolling back-

FIGURE 48
This pipe buggy can carry a log weighing up to 1000 pounds.
Sumner Manufacturing Company, Inc.

Figure 49
A pipe dolly can carry a log weighing up to 2,000 pounds.
Sumner Manufacturing Company, Inc.

wards down the ramp. The dolly should be pushed into the truck bed short-end first since it will not fit long-end first.

Logs are more easily loaded into a trailer since the ground clearance of most trailers is much lower than pickup truck beds. Trailers with attached ramps, the kind used by lawn services to load and unload lawn mowers, are even better. Having loaded both ways, I much prefer a trailer. Loading is faster, requires less time and energy and is safer.

The two pipe carriers pictured above can be rented for modest fees and used only as long as needed. Other than manually rolling the logs across the ground, these are the lowest cost skidding options I know of. There are other types of equipment beyond the pipe carriers, although not as large and expensive as the skidder used in commercial logging pictured in the last chapter (Figure 26). Lift attachments for tractors can be used to carry logs. At the next level, there are grapple rakes whose heavy curved tines can grasp and hold a few modest-sized logs at once. Then there are logging winch attachments for tractors that can lift and skid several large logs at once. There are forwarder trailers that can be towed by tractors. The trailers come with and without knuckleboom log loaders. These machines are feasible where space and ground conditions permit, which excludes them from use in most urban yards. If you do enough of this, one more option that would not cost much is to make your own log carrier.[4]

Do not overload the truck or the trailer. Too much weight in the bed shifts the weight distribution toward the back of the truck, which in turn, can dangerously interfere with steering. More than once I have pulled a log out of the bed of my pickup because the steering felt light and unresponsive. You could have a serious accident if your truck cannot respond adequately in an avoidance maneuver. At a minimum, overloading a trailer will cause excessive wear on the wheel bearings. Too much weight could cause a blowout or a tire rim or axle failure.

Whether you haul in a truck or trailer, you must tie the logs (or lumber) securely in place. Load shifts can cause the truck or trailer to move suddenly and dangerously while you are driving. Worse yet, shifts could end with the logs spilling out onto the road, creating a sudden and extreme hazard for other motorists. If the logs do not fit tightly into the bed or trailer, then they should be strapped together so that they will not roll from side to side. Start by placing the straps in the bed or trailer first and then loading the logs. Strapping the logs together is easier this way than if you load first and then try to pull the straps under the logs. Once the logs are strapped together, they should be securely tied down in the truck bed or trailer, especially if you are driving with the tailgate down or with the logs protruding beyond the rear edge of the trailer.

Check with your state's Department of Motor Vehicles for the maximum overhang and find out what must be tied to the end of the protruding material (for example, a red flag tied on the end of the overhanging logs). Accelerate slowly, otherwise, in a variation on the trick of jerking a tablecloth off a table while leaving the dishes in place, you may drive out from under the logs, dumping them in the street. I once dumped a load of poorly-secured lumber in an intersection because I accelerated too quickly at the green light. No damage was done, but this is embarrassing.

Species, log condition and size, accessibility, and hauling capacity are considerations that determine how much effort and expense I will incur to retrieve a log. In my part of the country, walnut, cherry, hard maple, pecan, oak, and Osage orange are woods that I prize the most and will try to retrieve. Ash, hickory and soft maple are second priorities. All others have to be readily accessible and easily transported to the mill. Exceptions are trees with burls, unusual grain, or trees that have historical or personal value. Each of us has to decide how much effort and expense we are willing to commit to particular logs. In the end, some potentially reclaimable trees are not worth the effort. Economists argue that the real cost—they call it the opportunity cost—of acquiring something is measured by the value of what we give up to get it. The opportunity cost of retrieving logs of lesser value is the loss of the more valuable logs we could have retrieved instead with the same effort. In addition, for woodworkers, the opportunity cost of reclaiming trees of lesser value is the greater value of what we could have done in our shops.

There is one other point that requires attention before we turn to cutting the logs into lumber. If the logs are not going to be cut right away, two steps must be taken to protect them from cracking and from insect and fungal damage. Whether at the site where they were felled, or where they are to be cut, the logs should not be left lying on the ground. The point at which they touch the soil is the entryway for boring insects and moisture that promotes fungal growth. Both will eventually destroy the logs' value as lumber. Simply roll the logs up on pieces of stone or brick so that there is an air space between them and the ground. Second, coat the ends of the logs with a sealant. I use a product called Anchorseal, a wax-based liquid that retards end-drying and related end-cracking. There are other products that do the same. If the logs are going to be stored for several months before cutting, especially over warm months, then you should keep them moist. Watering is not necessary during cold weather, especially when the temperature is at or below freezing, since drying stops below freezing.

So far, we have skidded, yarded and loaded logs for hauling to the mill. However, there is an alternative to hauling.

Bringing the Mill to the Logs

With a wide variety of portable sawmills available in or near every urban area in the country, an alternative to hauling the logs to a mill is paying the sawyer to come to the logs and cut them there. There is a lot to be said for this:

First, you avoid the time and expense incurred in hauling the logs. If you do not have a large enough truck or trailer, then there is the additional expense of renting one or both. Even with the right truck for the job, you still have to load the logs. This means either paying laborers or imposing on your friends for help. You might be surprised to learn that, after the novelty of it wears off, most of your friends do not consider loading logs to be an entertaining way to spend a day.

Second, you avoid loading-related injuries, principally mashed fingers and toes and strained back, shoulder and arm muscles.

Third, by cutting on-site, you completely avoid transportation mishaps, like the logs shifting or even falling out of the bed of the truck while you are underway.

Fourth, if you plan to store the lumber on-site in your garage, basement, or outbuilding, then you can simply sticker and stack the lumber as the logs are milled. Friends may be more agreeable about helping when they can watch the sawmill in action and are not required to lift more than one board at a time. Also, injuries are less likely when carrying individual boards than when loading logs.

Fifth, even if you do not plan to store the lumber on-site, hauling several boards at a time is a lot easier and safer than transporting the entire log. And you can use smaller trucks to do the hauling.

From my experience, there are only two disadvantages to having the mill come to the logs and they are relatively minor:

In the first place, you will probably pay the sawyer for travel time to and from the site. And in the second place, there is also the need to clean up sawdust, slabs of unusable wood, and bark from the day's cutting. If you place a large sheet of plastic along the cutting path where sawdust accumulates, then cleanup goes much quicker.

Overall, the cost of having the mill brought to your logs will be much less than the time you would spend and the expense you would incur in loading and then hauling logs to the mill.

Transportable Sawmills

Options

Even though you pay the sawyer to do the sawing, you can and should be a knowledgeable observer, able to ask intelligent questions and to make informed choices about the way you want the logs sawn. The first choice is the type of mill to select: chainsaw mills, transportable circular sawmills, and transportable thin-kerf band mills.

Chainsaw Mills

Chain saws can be fitted with special chains, longer bars, cutting frames, and manual cranks that enable the operator to saw logs.[5] Less expensive attachments allow the saw to ride and cut along the top of the log while heavier frames can accommodate both the saw and small logs. Although chainsaw motors pack a lot of power for their size and weight, they cannot match the sawing pace of the more powerful circular mills and band saw mills. In addition, the total kerf—the swath cut in the log or board by the blade itself—is at least 1/4 inch, comparable to circular saw blades but much wider than band saw blades. While the chainsaw option is less expensive to purchase, it produces more sawdust and is less productive than either of the other two.

Circular Saw Mills

Transportable circular saw mills are a second option.[6] Large, more fully equipped models come with powered log-lifting, turning and positioning arms, and dogs (stops that hold the log in place while it is being sawn). Since the blade in most of the larger models is fixed, a powered feed table is used to move the log through the blade for the cut. On larger models, the blades are roughly 4 feet in diameter. Even so, their depth-of-cut is slightly less than half their diameters because of the blade eye and arbor (the hole in the middle of the blade and the powered shaft that fits through it). For example, a 1,200 mm blade (about 47 inches), has a depth-of-cut around 500 mm (about 19 inches). Some manufacturers provide a top saw option—another blade positioned above and slightly forward of the lower blade) that can increase the total cut another foot or so. Smaller models use smaller blades that cut to a depth of about a foot. The total kerf of circular saw blades is at least 3/16 of an inch; even more if the blade wobbles, or if the teeth are misaligned or dull. A cut this wide means that, over a large number of logs, more wood becomes sawdust than lumber when compared to thin-kerf band saw mills. Because more wood has been removed between two adjacent boards, precision book matching is a bit more difficult as well. In addition, these blades are expensive (as much as several thousand dollars for the larger ones). A circular saw operator might be unwilling to cut urban trees unless you agree to cover the replacement cost of an irreparably damaged blade.

Thin-Kerf Band Saw Mills

The third and most commonly used option is the thin-kerf band saw mill [7] .The blade used on these mills is a continuous steel band with uniformly spaced hooked teeth on one edge. Blades are 1-1/4 and 1-1/2 inches wide, and 0.4 to 0.5 inch thick. The kerf is about 1/16 of an inch. A well-tuned mill is capable of sawing near-veneer-thin slides of wood: my son and I cut ten successive slices slightly less than 1/8 inch each from a bur oak log to make the table top shown in Figure 62. The blades are inexpensive (about $20) and can be re-sharpened several times, either with the mill-owner's own sharpening equipment (which several mill manufacturers sell) or by the manufacturer.

In operation, the blade rotates on two wheels: one wheel is driven by either an electric motor or a gas or diesel engine (ranging from 15 to 40 horsepower). The log is held in a fixed position on the sawmill carriage. The frame that holds the engine, clutch, wheels and blade—sometimes referred to as the mast—passes over the log while the turning blade cuts through the log. Though there are a few mills that cut in both directions, most cut only in one direction. After the cut, the clutch is disengaged and the blade stops turning. The board that has just been cut is removed from the top of the log, and the mast moves back to the other end of the mill so that it is positioned for the next cut. The mast that holds the motor, clutch, wheels, blade, controls, and even a riding seat on some models, either rides on top of two parallel rails that form the bed of the mill or it rides on just one of the bed rails.

The presumed advantage of the parallel rail configuration is greater mast rigidity and, therefore, cuts of consistent thickness. An important disadvantage is that there is an absolute limit to the size of the log that will pass between the upright posts of the mast. A log wider than the distance

FIGURE 50
Author cuts a cherry log on a medium-size mill, Wood-Mizer's LT-15 with a dual-post mast.
Photo by Patricia Sherrill

FIGURE 51
Author uses a large mill with a cantilevered mast to cut a walnut log wider than this mill could cut down the middle. By turning and trimming, this rig can cut boards up to 32 inches wide.
Photo by Patricia Sherrill

between the posts cannot be cut because it just will not fit.

By contrast, the cantilevered mast can accommodate larger logs because there is only one post holding the mast, leaving the other side open. Whatever the maximum cutting width of this type of mill (twenty-eight to thirty-six inches is typical), it can cut wider logs by trimming and turning until the squared up log (called a cant) has been trimmed down to the width the mill can accommodate. Because of the way they are constructed, these mills will cut true even when they are not perfectly level to the ground, or when one of the outriggers sinks into the ground under the weight of a heavy log. The presumed disadvantage of the cantilevered mast is that it is not as rigid as its dual-post counterpart. If the mast sags slightly toward the open side, then the blade will not cut parallel to the bed. The last board will not be of uniform thickness but will, on-end, look more wedge-shaped. If the mast sags and is loose, then none of the cuts will be of the same uniform thickness. I have used two versions of the cantilevered mast and encountered no uneven cutting with either one.

The medium-size Wood-Mizer® LT-15 pictured in Figure 50 is an excellent entry-level machine that costs from $5,000 to $6,000 dollars, depending on how you accessorize it. It will cut logs up to, but absolutely not beyond, 26 inches in diameter. I have cut larger logs, but only by first trimming the excess off with a chain saw. Chainsaw trimming is worthwhile for logs that are just a few inches over the limit. Going much beyond this means you are cutting scraps instead of lumber and are also cutting the best quality wood from the log. In addition, chainsaw trimming is time-consuming. Like any tool, this saw works best when you stay within its limits.

Mills that sit directly on the ground must be level, and securely positioned. The mill shown below in Figure 50 is leveled by eight frame-adjustment bolts lowered on to footplates that sit on the ground. Being out of level will place the frame under extra stress and cause the mill to rock as the mast is moved back and forth. Once it is level, loading ramps attach to the side of the mill. The ramps come with spring-activated stops that flatten only for logs being rolled up on the bed. The stops prevent the log from rolling backwards. The mill also comes with a leveling wedge (called taper sets) to raise the narrower end of the log so that the center of the log is parallel to the saw bed.

With the blade running, the log is sawn by turning a crank that advances the mast along rails, and the blade through the log. Cutting speed on larger mills is variable, but once set is automatically maintained by the machine's power system. One advantage of the manually operated mill is that I control the cut rate, so I can slow down if I hear the motor begin to labor or if I hear a noise I do not like, such as the blade touching something embedded in the wood. Manual cutting is slower than automatic cutting; however, I can respond more quickly when cutting manually than I can when cutting automatically. This is an important advantage when cutting urban logs that might contain something the tree itself did not grow. On the other hand, larger machines with hydraulic loading and cutting are much more productive.

In my experience, the large mills are dream machines (as in Figure 51). They can be fully outfitted with electric and hydraulic systems that load, turn and level large logs. Mills outfitted with set works can automatically saw boards to one or more specific thicknesses and can be set for repeat cutting. A board off-loader will pull each board to the back of the mill as the mast returns to its starting position after each cut. Of course, these machines cost a lot more, require at least a full-size pickup truck for towing, and must have running lights and brake lights connected to the towing truck as well as electric brakes that work with the truck's brakes. And, the mill should be covered by liability insurance against damage while being towed or used on-site.

Sawing Costs

Pricing Options

Mill owners will charge either by the hour or day, or by the number of board feet cut. They might also cut for a lower fee plus a share of the lumber, or no fee plus a larger share of the lumber. Most will charge full or partial fees; very few are likely to cut only for a share since this means that, in order to recoup their expenses, they will have to sell the lumber.

You can expect to be charged an hourly rate of $40 to $60 plus travel time. There will probably be a minimum charge of two hours, perhaps more. While I would expect to pay for travel, I would not expect to pay the same hourly rate for travel time as for cutting time.

The flat daily rate for an 8-hour day could run from $350 to $500. Since this will include travel, setup, blade changes, and other minor adjustments, you will end up with 5 to 6 hours of actual cutting time.

The advantage of a flat hourly or daily rate is that the sawyer gets paid a fixed amount regardless of how many board feet are cut. This arrangement will work well for you if a small number of logs are to be sawn into many boards by an experienced operator. This would not be as advantageous if the operator is inexperienced and slower. Hence, you should ask how long the operator has been sawing and, specifically, how many board feet you can expect per hour, or per day, under ordinary working conditions.

The most common pricing is by the board foot (defined on the facing page). This rate can range from 10 cents to 40 cents per board foot depending on the species, log size and condition, the number of logs, the way the logs are to be sawn, and the degree of competition among sawyers in your area. Since shorter logs take about the same time and effort to saw as longer ones, but yield fewer board feet and lower revenue, some operators feel they do not make enough on short logs to justify charging the same rate. Hence, they will either charge more for short logs (3 to 6 feet in length) or will charge based on a minimum length (6 to 8 feet). Like the standard rate per board foot, whether you pay extra for shorter logs will depend on the degree of competition among owners.

When using board foot pricing, the incentive for you is to have the sawyer cut as many boards of lesser thickness as possible (for example, all 1-1/8 inch thick). For the operator, the incentive is to cut fewer boards of greater thickness. Part of negotiating the cutting fee is determining whether there are extra charges for thin or wide cuts. Some sawyers might charge extra for cutting boards one and two inches thick, and they may charge a premium for wide boards that require a slower feed rate. One very experienced and skillful sawyer I know charges 40 cents per board foot but is willing to cut any one of several ways, thin or thick, and wide or narrow, from any species. I save the most demanding sawing jobs for him.

Sawing Costs and Log Scaling

If you agree to pay by the board foot, the total sawing cost is easily estimated in advance by multiplying the board foot price by the estimated number of board feet the logs are expected to yield. Even when you agree to pay by the hour or day, you will still want to know in advance approximately how much lumber you can expect. Dividing the agreed-upon or expected payment by the estimated number of board feet gives you the estimated price per board foot. After the job is completed, you can tally the amount sawn and compare the actual to the estimated price per board foot. Unit pricing reduces bids and estimates to comparable measures that, in turn, enable you to choose the offer that meets your budget as well as the required amount of lumber for projects or buyers.

Recall that a board foot is a volumetric measure equal to a board that is 1 inch thick by 12 inches wide by 12 inches long, or 144 cubic inches. A board 1 inch thick, 2 feet wide, and 6 feet long contains 12 board feet: 1 inch times 24 inches (two feet) times 72 inches (6 feet) equals 1,728 cubic inches. Divided by 144 cubic inches

CALCULATING BOARD FEET

1 inch thick x 12 inches wide x 12 inches long = 144 cubic inches = 1 board foot

1 inch thick x 24 inches (2 feet) wide x 72 inches (6 feet) long = 1,728 cubic inches

1,728 cubic inches / 144 cubic inches/board foot = 12 board feet

or

1 inch x 24 inches wide x 6 feet long = 144 / 12 = 12 board feet

12 board feet x $0.25/board foot = $3.00/board

100 boards @ $3.00/board = $300.

(1 board foot), this board contains 12 board feet, and 100 of these boards would contain 1,200 board feet.

A quicker method is to multiply the thickness in inches by the width in inches by the length in feet (ignore the difference in units for the moment) and then divide by 12. This is faster but the other method clearly illustrates the numerical meaning of a board foot. Either way, at 25 cents per board foot, for example, the sawing cost for the 12 board feet would be $3 (25 cents per board foot times 12 board feet) or $300 for the 1,200 board feet.

Estimates of the board-foot content of logs must be made to calculate the expected yield and sawing cost. Estimating requires a log rule and log scaling. Log rules—and there are many of them—are usually based on one of three specific methods of estimating the board foot content of logs. Rules have been devised either from equations based on numeric assumptions about such factors as kerf and taper, or geometrically, using diagrams of the ends of boards imposed on cross-sections of logs. In addition, some sawyers use mill-specific rules-of-thumb, which reflect their own equipment and experience. Log scaling is the application of a log rule.[8] Tables of estimates for common diameters and log lengths are readily available for the most widely-used rules. A log or scaling stick, calibrated with one or more rules, is often used by professionals to speed up the estimation process.

Constructing a standardized rule that provides consistently accurate estimates over logs of all combinations of diameters and lengths would be straightforward if all logs were perfectly cylindrical like pipes, straight (no sweep), and were free of

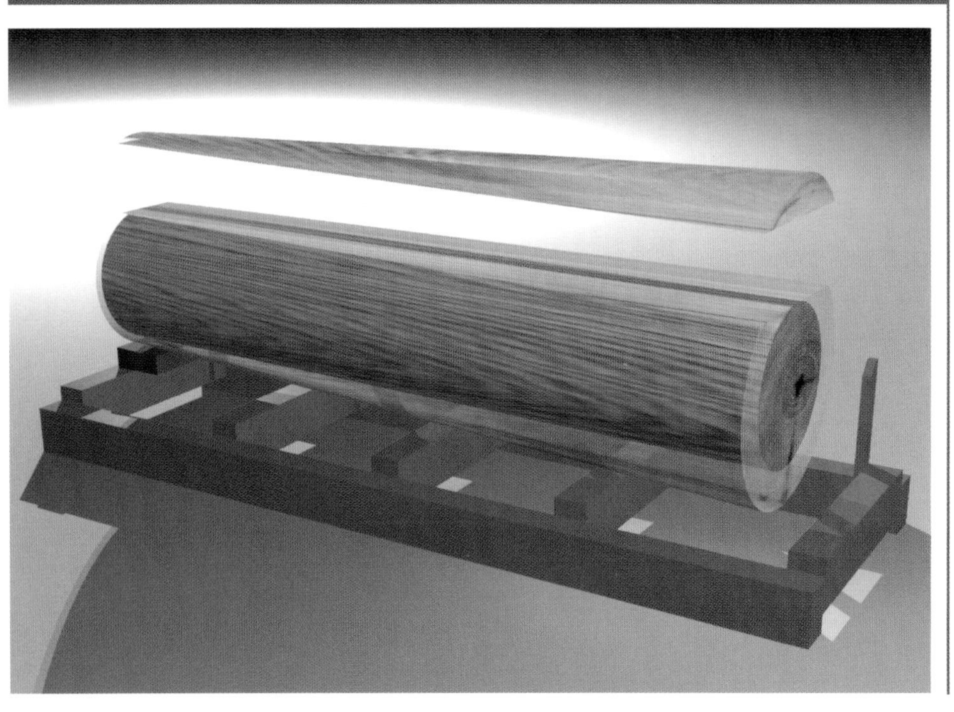

FIGURE 52
As shown from left to right above, the scaling cylinder is the circular small end of a log, inside the bark, evenly projected to the large end. For rules that do not incorporate taper into board-foot estimates, such as the Doyle and Scribner scales, this cylinder defines what will become lumber, from inside the cylinder and what, as slabs, will become waste.
Computer-generated image by Carey Sherrill

defects from the center to the bark. Such a rule would presume a specific kerf (such as 1/4 inch), type of saw and the sawyer's skill, board thickness, minimum width, maximum length, and slab thickness (as illustrated in Figure 52, slabs are the waste wood, rounded on one side, that comes from sawing square and rectangular boards from round logs).

Unfortunately, real logs are tapered, not cylindrical, often contain sweep, and have defects that can substantially reduce their board-foot yield. Typical defects are rot, ring shake, surface cracks, crotch wood where two or more stems converge, damaged ends, lightning scars, or deeply embedded objects that must be cut from the log. Appearance defects such as stain are not relevant since they have no impact on the board-foot yield of the log, only on how the boards sawn from it are graded and used.

Kerf varies widely depending on the type and condition of the blade, and the power and uniform action of the sawing mechanism. Sawing skills vary widely among sawyers, a factor that also influences yield but is difficult to incorporate into any widely-applicable rule. As discussed below, there is a basic way to saw a log that usually yields the largest amount of high-quality lumber. Sawyers who do not follow this method, or simply cut the log in the position it happened to land on the mill, will produce less lumber of lower quality than predicted by a log rule.

More precisely, rules based on equations and on geometric drawings rest on exact assumptions about how the log will be sawn. To the degree that actual logs and mills do not conform to one or more of these assumptions, log-scaling estimates and the actual amount of board feet sawn (called the mill-tally) will be unequal, sometimes by significant amounts. Overrun is the excess by which the mill-tally exceeds the log-rule estimate. Underrun is the opposite: the amount the mill-tally falls below the estimate. Neither of these is desirable, especially when the difference is substantial. An overrun means more lumber but at a cost that might exceed your budget. An underrun means a budget savings but with less lumber than anticipated for particular projects or buyers. There is no virtue to either inaccuracy, at least for the buyer.

By one count, there are more than 95 recognized rules, having about 185 different names, in use in the U.S. and Canada.[9] And even these rules have different local versions. The Doyle, Scribner Decimal C, and the International 1/4-Inch rules are the three most widely used in the U.S. However, because their assumptions differ, these three rules do not provide the same board-foot estimates for a given log. In addition, the Doyle and Scribner C are not consistent over the usual range of log diameters and lengths.

Developed by Edward Doyle around 1825, the Doyle rule is an equation mainly based on assumptions about kerf, slab size, and taper. Of the three, it is considered the least accurate and least consistent over different lengths and diameters. Because it is based on an overly generous assumption about slab size, a 25 percent reduction for kerf, and because it ignores taper, this rule underestimates the number of board feet in small logs and overestimates the number in large ones. Overrun will occur and sawing costs will be greater than expected for small logs. Just the opposite will happen with large logs: underrun will lead to less-than-expected board-foot yield and lower costs. This rule is most accurate for logs between 2 and 3 feet in diameter where both yield and costs should be about what is expected. As an example, in Table 3.1 below, by the Doyle scale a 16-foot log 8 inches in diameter contains 16 board feet. By contrast, the Scribner Decimal C (Table 3.) shows 30 board feet (add a zero to each number in the table) while by the International 1/4-Inch rule (Table 4.) estimates the log at 40 board feet. The Doyle is used more in the southeastern U.S. and in some of the mid-western states.

Widely used reference tables provide estimates for logs up to 30 inches in diameter and 16 feet in length. The following Doyle formula can be used to estimate the number of board feet in logs over 30 inches in diameter, 16 feet in length, or both:

$$\text{bd. ft.} = ((D - 4)/4)^2 \times L$$

where D = log diameter in inches and L = log length in feet.

For a log 20 inches in diameter and 6 feet in length:

$$\text{bd. ft.} = ((20 - 4)/4)^2 \times 6 = (16/4)^2 \times 6 = 16 \times 6 = \underline{96}$$

For a log 40 inches in diameter and 6 feet in length:

$$\text{bd. ft.} = ((40 - 4)/4)^2 \times 6 = (36/4)^2 \times 6 = 81 \times 6 = \underline{486}$$

J. M. Scribner developed the Scribner rule around 1849. Unlike the Doyle equation-based rule, the Scribner table for logs of different lengths was computed from scaled diagrams of board ends, drawn within circles representing log cylinders of common diameters. This rule is more accurate than the Doyle but less than the International 1/4-Inch rule. The Scribner consistently underestimates board-foot content, in part because it, like the Doyle rule, does not include taper. Underestimation increases as log size increases, hence overrun will consistently occur with this rule, increasing along with sawing costs for progressively larger logs. The latest version, the Scribner Decimal C given in Table 2, is used mainly in the western United States for softwoods (it rounds estimates to the nearest 10 board feet and drops the zero). A slightly different version is often used for western hardwoods, and by the USDA Forest Service.

Since the Scribner rule is based on scaled diagrams, there is no formula to calculate board-foot estimates for logs over 30 inches in diameter or 16 feet in length, other than what would be based on the diagrams themselves.

In *Forest Measurements*, the standard text on measuring timber, Avery and Burkhart express their low regard for the combination of these two scales: "... the more erratic attributes of the Doyle and Scribner log rules are combined to form a diabolical yardstick called the Doyle-Scribner log rule."[10] The two together yield the lowest board-foot estimates. The uninformed log seller would lose money to a clever but not especially honest buyer who uses one of these "diabolical" sticks. You could find yourself paying quite a bit more than you were led to believe when this particular yardstick is used by the sawyer to bid your job.

The International 1/4-Inch rule, first formulated by Judson Clark in 1900 as a 1/8-inch kerf rule and subsequently altered to the larger 1/4 -inch kerf in 1917, is considered to be the most accurate and consistent of the three. This rule is superior in both ways to the other two because it is one of the few rules that have a built-in allowance for the tapered shape of the log (half an inch for every 4-foot section). By this rule, the imaginary scaling cylinder of a log is approximately conical in shape, not cylindrical. It is also based on more precise estimates of loss, due to the kerf of the saw blade and also to slabbing. Both overrun and underrun, thus board-foot yield and

TABLE 2

Doyle Log Scale

	Length of Log (feet)					
Diameter	6	8	10	12	14	16
(Inches)	Contents in Board Feet (tens)					
8	6	8	10	12	14	16
9	9	13	16	19	22	25
10	14	18	23	27	32	36
11	18	25	31	37	43	49
12	24	32	40	48	56	64
13	30	41	51	61	71	81
14	38	50	63	75	88	100
15	45	61	76	91	106	121
16	54	72	90	108	126	144
17	63	85	106	127	148	169
18	74	98	123	147	172	196
19	84	113	141	169	197	225
20	96	128	160	192	224	256
21	108	145	181	217	253	289
22	122	162	203	243	284	324
23	135	181	226	271	316	361
24	150	200	250	300	350	400
25	165	221	276	331	386	441
26	182	242	303	363	424	484
27	198	265	331	397	463	529
28	216	288	360	432	504	576
29	234	313	391	469	547	625
30	254	338	423	507	592	676

Source: Burl S. Ashley. October, 2001. Reference Handbook for Foresters, NA-FR-15, U.S. Department of Agriculture, Forest Service, State and Private Forestry, Northeastern Area, p. 6.

sawing costs, should about equal estimates. The International rule is widely used in the eastern U.S. The USDA Forest Service uses it and a variation, the International 1/4-Inch Decimal rule, that rounds estimates to the nearest ten board feet.

While the International rule is more accurate than the Doyle, the board-foot estimates for logs more than 30 inches in diameter and 16 feet in length are somewhat more complicated as well. Calculations are done separately for each 4-foot section, then summed for the total number of board feet. The basic formula is:

$$\text{bd. ft.} = 0.905((0.22(D)^2) - 0.71D)$$

TABLE 3
Scribner Decimal C Scale

	Length of Log (feet)					
Diameter	6	8	10	12	14	16
(Inches)	Contents in Board Feet (tens)					
8	1	1	2	2	2	3
9	1	2	3	3	3	4
10	2	3	3	3	4	6
11	2	3	4	4	5	7
12	3	4	5	6	7	8
13	4	5	6	7	8	10
14	4	6	7	9	10	11
15	5	7	9	11	12	14
16	6	8	10	12	14	16
17	7	9	12	14	16	18
18	8	11	13	16	19	21
19	9	12	15	18	21	24
20	11	14	17	21	24	28
21	12	15	19	23	27	30
22	13	17	21	25	29	33
23	14	19	23	28	33	38
24	15	21	25	30	35	40
25	17	23	29	34	40	46
26	19	26	31	37	44	50
27	21	27	34	41	48	55
28	22	29	36	44	51	58
29	23	31	38	46	53	61
30	26	33	41	49	57	66

Source: Reference Handbook for Foresters, p. 5.

where D equals log diameter in inches at the narrowest end.

To account for taper, D is increased by 1/2 inch for every 4-foot section starting with the second section. For a 16-foot log 20 inches in diameter, the number of board feet in each of the four 4-foot sections, starting with the narrowest section, is:

bd. ft. (1) = $0.905((0.22(20 + 0)^2) - 0.71(20 + 0)) = \underline{67}$

bd. ft. (2) = $0.905((0.22(20 + 1/2)^2) - 0.71(20 + 1/2)) = \underline{71}$

bd. ft. (3) = $0.905((0.22(20 + 1)^2) - 0.71(20 + 1)) = \underline{74}$

bd. ft. (4) = $0.905((0.22(20 + 1\text{-}1/2)^2) - 0.71(20 + 1\text{-}1/2)^2) = \underline{78}$

TABLE 4
International 1/4-Inch Scale

	Length of Log (feet)					
Diameter	6	8	10	12	14	16
(Inches)	Contents in Board Feet					
6	5	10	10	15	15	20
7	10	10	15	20	25	30
8	10	15	20	25	35	40
9	15	20	30	35	45	50
10	20	30	35	45	55	65
11	25	35	45	55	70	80
12	30	45	55	70	85	95
13	40	55	70	85	100	115
14	45	65	80	100	115	135
15	55	75	95	115	135	160
16	60	85	110	130	155	180
17	70	95	125	150	180	205
18	80	110	140	170	200	230
19	90	125	155	190	225	260
20	100	135	175	210	250	290
21	115	155	195	235	280	320
22	125	170	215	260	305	355
23	140	205	255	310	370	425
24	150	205	255	310	370	425
25	165	220	280	340	400	460
26	180	240	305	370	435	500
27	195	260	330	400	470	540
28	210	280	355	430	510	585
29	225	305	385	465	545	630
30	245	325	410	495	585	675

Source: Reference Handbook for Foresters, p. 4.

Total bd. ft. in 16-foot log, 20 inches in diameter = 290

For a 16-foot log 40 inches in diameter, the number of board feet in each of the four 4-foot sections is:

bd. ft. (1) = $0.905((0.22(40 + 0)^2) - 0.71(40 + 0)) = 294$
bd. ft. (2) = $0.905((0.22(40 + 1/2)^2) - 0.71(40 + 1/2)) = 301$
bd. ft. (3) = $0.905((0.22(40 + 1)^2) - 0.71(40 + 1)) = 308$
bd. ft. (4) = $0.905((0.22(40 + 1\text{-}1/2)^2) - 0.71(40 + 11\text{-}1/2)) = 316$
Total bd. ft. in 16-foot log, 40 inches in diameter = 1,219

These calculations apply to log lengths that are multiples of the 4-foot sections. Calculating the board feet in logs that are not multiples of 4 is even more complicated. To save time, calculate the nearest multiple of 4 above and below and split the difference. For example, there are 325 board feet in an 8-foot log that is 30 inches in diameter, and 495 board feet in a 12-foot log. Splitting the difference of 175, then adding 85 to 325 gives you 410 board feet, the exact estimate for a 10-foot log. As you can see in Table 4, this works progressively better for larger diameters and longer lengths.

Advance Arrangements Reduce Costs

Whether you are paying by the hour, day, or board foot, you want the sawyer 's time to be spent cutting logs. Only this way will you get the most lumber for what you pay. This means that you want to minimize set-up time by selecting a site that is easily accessible and level. As part of the cutting arrangement, you should give the sawyer a complete and accurate description of the cutting site, with special attention to obstructions and steep terrain.

On one cutting job I did, the property owner told me that the logs were right next to the barn where he would store the lumber. He said that the space was flat and clear. What he did not tell me was that to reach this ideal site, I had to drive down a steep hill, across a narrow creek bed, and up another equally steep hill. Going downhill was no problem. Crossing the narrow creek bed was a problem. The truck and the mill, together about 30 feet long, just barely fit in the narrow valley between the two steep hillsides. The other hill was too steep for my rear-wheel-drive truck to pull alone. Fortunately, the owner had a tractor and, between the two vehicles, we got the mill to the top of the hill. Getting into place and setting up consumed close to two hours—time that could have been spent cutting.

The sawyer will want to know the species, size, quantity and condition of the logs, as well as whether they are fresh-cut or have been down for some time. Clean and freshly-cut poplar logs 2 feet in diameter and 8 feet long can be sawn faster than oak the same size and in the same condition. For all species and sizes, sawing will go more slowly on logs that have been on the ground for many months, as it will for those that have been dragged across the ground, picking up mud and small stones along the way. Small rocks and other hard debris not only dull the blades but the also cause wavy and uneven surfaces on the boards.

If you look back at Figures 50 and 51, from the perspective of the camera, the blade cuts from the right to the left side of the log. If the blade snags a small pebble on the right side it will drag it through the log and out the left side. This dulls the blade, unnecessarily strains the mill, and produces an uneven board surface. Cleaning the logs in advance will improve the quality of the lumber, and will minimize downtime for blade changes. Larger mills come with debarkers which, on the larger Wood-Mizer® models, are essentially small circular saws. Equipped with carbide tipped blade, the debarker removes bark and debris and clears a clean path ahead of the blade. This device makes blades last longer and board surfaces smoother.

Clearly mark all surface defects you can spot, such as knots, holes, decay, unnatural bulges (where the tree might have grown over something), and other defects that can mar the surface of the boards. Marking these places will enable a skilled sawyer to cut the log in a way that minimizes defects in board faces.

You should also tell the sawyer whether any of the logs have metal or other foreign material in them (these logs should have been marked with spray paint during limbing and bucking). In advance, cut all that you can out of the logs. Since you might still miss something more deeply embedded, place these logs, as well as others you suspect might contain foreign material, last in line to cut. Blades are quickly dulled by these encounters and have to be replaced, reducing the time available for sawing. By sawing your best logs first, you can keep blade changes to a minimum and get all of the best lumber first. Save the worst for last so that, if there are problems, you can elect not to cut logs that would yield less anyway.

Most sawyers will not charge for routine blade changes: this should be built into their fees. Many will charge for blades irreparably damaged by metal or concrete, however, so you need to determine in advance whether a particular log is worth the risk. I expect a charge of about $30 for a ruined blade. So, the question is whether a log is worth the standard cutting fee plus additional charges for damaged blades. Size, species and quality are the factors that influence my decision. For example, I would take the chance for a large black walnut log in otherwise good condition, and would not do so for a medium-size soft maple.

Have the logs lined up parallel to the place where the mill will be parked. This way, the logs can be quickly rolled up the ramps onto the mill or loaded by the hydraulic lift. At least two off-bearers are needed to lift the boards from a medium-size mill and at least four are needed for large hydraulic mills. The sawyer pauses after each cut while the boards are off-loaded, and then moves the mast back to the starting position for the next cut. The key to maximizing productivity—that is, getting the most lumber for the time and money spent—is minimizing the pause between cuts. Two off-bearers are somewhat faster than one but, more importantly, given the weight of wet boards, two are needed to safely lift each board. Hours spent reaching over the mill bed to grab a board will produce shoulder and back aches. Having more than four volunteers will provide each of them with a break while allowing the sawyer to work continuously.

If the mill and operator are cutting quickly, the off-loaders can simply stack one board on another and wait until later to load them for hauling or to sticker them for drying. This is important: within the day the sawing is completed, you should sticker and stack the lumber for drying, especially during warm weather. Since fungus will quickly begin to grow on the faces of wet boards stacked one on top of another, there should be little delay in separating them into layers with air spaces between each layer. I describe how this is done in the last section of this chapter.

Ask if the mill is equipped with a de-boarder, a device that on the large Wood-Mizer® band mills is a flange that drops down over the end of the board and pulls it back off of the top of the log as the mast moves back to the starting position for the next cut. The de-boarder eliminates the pause between cuts needed to manually pull the board from the top of the log. It also eliminates the need for off-loaders to reach across the mill bed to reach a board. Instead, they can wait at the end of the mill and simply grab the board as it is pushed off the end of the log where the next cut begins. The de-boarder increases productivity and adds to the margin of safety. On a large hydraulic mill without a de-boarder, I watched an inexperienced operator start the mast moving back to the starting position for the next cut without first checking to see if anyone was behind him. One of the off-loaders was in the path of the mast trying to get a better grip on the board he had just pulled off of the log. The mast knocked the hat off of his head as it passed him. Fortunately, he was partially bent over so that it was the hat that got knocked off, not his head.

While these mills are very safe when operated properly—I'm convinced by my own experience that you would have to work at getting cut—injuries are still possible. In addition to being struck by a moving mast, fingers, arms, and feet can be injured while manually rolling a log onto the mill or positioning it on the mill bed. This usually happens when several people are simultaneously pushing and pulling a log into position on the mill bed and fingers get caught between the log and the metal stops (referred to as dogs) that hold the log in place while it is being sawn. Coordination by the mill operator, as opposed to everyone trying something different at once, helps reduce these kinds of injuries.

In addition, cant hooks (see Figure 45 in the previous chapter) make the task of maneuvering a log easier on the back, and safer for hands and feet. These types of injuries are much less likely on mills equipped with powered lifts that can load and then rotate logs into position. Cutting is a repetitive task which, after several hours, can lead to lapses in attention. Work breaks are important, not only for the off-loaders' backs and shoulders, but for the sawyer as well. Taking time to rest and refocus is also productive because it reduces injuries that cost time and money, and that impose greatly on your volunteers.

The mill itself is relatively easy and safe to operate. However, the weather can create unsafe operating conditions that are often overlooked. Being struck by lightning is not as rare as the cliché implies: it happens to more people than just golfers in Florida. According to the U.S. Weather Service, lightning kills an average of 93 people every year, more than tornadoes or hurricanes, and injures 300.[11] Lightning is second only to floods as a cause of storm-related deaths. While 10 percent of those struck die, the 90 percent who survive are left with chronic health problems, most often severe neurological disabilities.[12]

As a rule, if you can see lightning and within 30 seconds hear the thunder that follows, you are close enough to be struck, even if the sky overhead is clear. When thunder follows lightning by 30 seconds or less, everyone working at the mill should immediately stop what they are doing and go to safe shelter. Immediacy counts, since injuries often occur as people are headed for shelters, suggesting that if they had just started a few minutes sooner, no one would have been struck. Well-constructed buildings (homes, commercial buildings, and barns) provide adequate shelter, as do vehicles (cars with enclosed roofs, and trucks). The sawmill is not likely to be safe because it is open. Likewise, stay away from trees. Lightning can spread from the point where it strikes, so you can be injured even though you are dozens of feet from the strike point. Finally, do not resume cutting until the time lapse between lightning and thunder exceeds 30 seconds.

Sawing Logs into Lumber

Sawing for Grade

Before sawing begins, you and the sawyer should review your judgment of log quality and cutting order. A sawyer who is experienced and knowledgeable might see defects you missed and recommend a different order. Recall that defects arise from embedded limb stumps, something embedded in the tree that did not grow there, holes in the side of the log, or decay visible from either end. Also, defects are likely where there are breaks or bulges in the bark, where bark is missing, or any place where the bark looks as if it has healed over an injury or a lightning strike.

A skilled sawyer should be able to briefly describe how each log will be cut to minimize defect in board faces and thereby maximize the quality of the lumber. This is referred to in the lumber industry as sawing for grade: that is, sawing from logs in a way that yields boards of the highest quality as judged by commercial standards such as those established by the National Hardwood Lumber Association (NHLA) described in the section below.

Sawing logs for grade—to get the best quality wood from each log—is essential to the whole idea of harvesting urban timber. Making the best use of all urban trees requires not only that we treat them as a resource, but also that we get the best lumber we can from each tree, otherwise our efforts are diminished or even wasted. For this reason, understanding the basic prin-

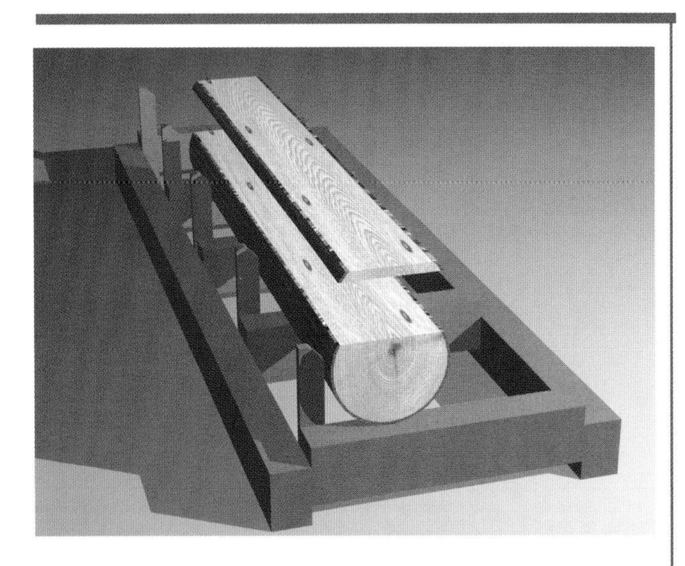

FIGURE 53
Notice that the first cut made from this log has placed most of the defects at the edges of the board leaving it with a nearly clear face.
Computer-generated images by Carey Sherrill

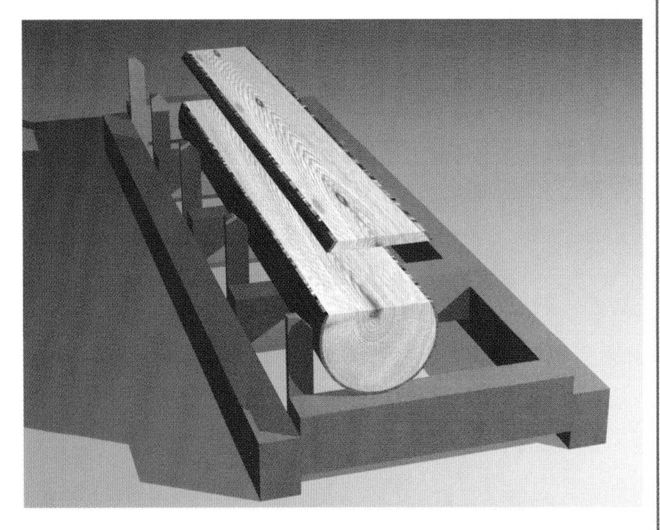

FIGURE 54
Here the log has been sawn without regard to the location of a large knothole. The first and subsequent boards cut from this log will contain a major but avoidable defect that will lower their grade and value.

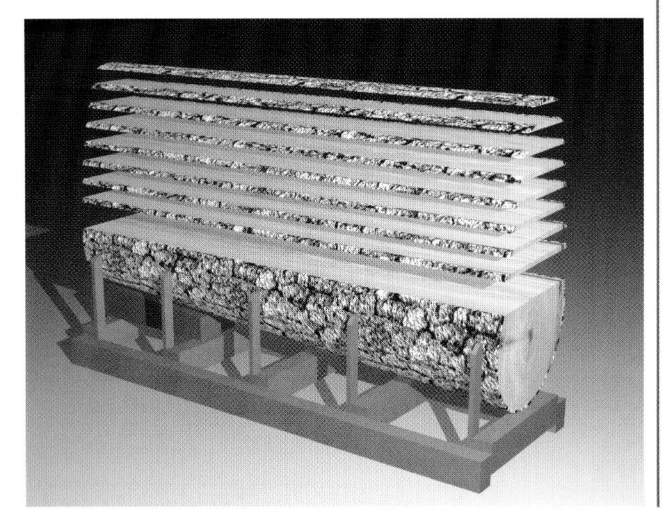

FIGURE 55
Opposite good faces are sawn by first elevating the narrower end (left above) of the log so that the top of the log is lengthwise parallel to the mill bed. Boards are sawn until the grade drops.

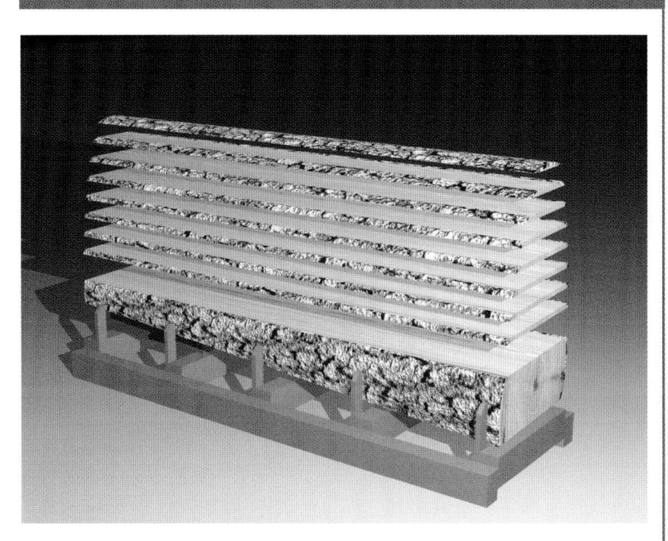

FIGURE 56
The log is rotated a hundred and eighty degrees so that the other good face is up. The taper remains in place and boards are again sawn until the grade drops.

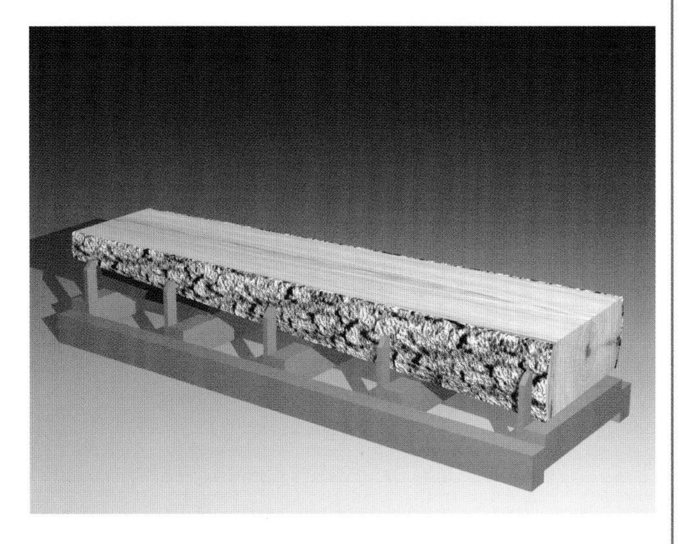

FIGURE 57
Cutting opposite good faces parallel to the mill bed leaves a wedge-shaped two-sided cant.

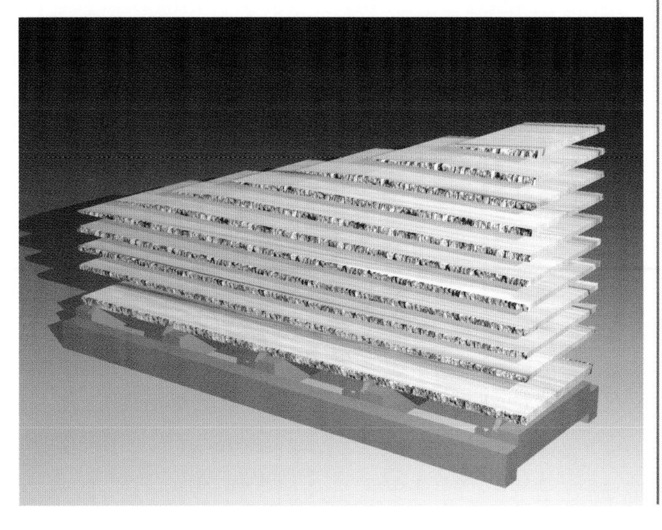

FIGURE 58
The cant can be squared up by sawing to the vicinity of the pith. This produces several short boards with tapered ends. The squared cant would then be rotated and sawn again to the vicinity of the pith. This boxes the heart or pith in the middle of a thin wedge-shaped board that would be discarded. Or, as shown above, the dogs can be lowered and the cant can be sawn down to the bed.

ciples of cutting for grade is essential.

Described immediately below are basic principles of grade-sawing straight logs with surface defects into cants. Cants are logs that have been sawn so that they have at least two parallel faces; when completed, the cant has four flat sides, each of which is perpendicular to its two adjacent faces. Boards are then sawn from the cant faces. Sawing to compensate for sweep (curvature in the log) is also described. Grade sawing for more complicated defects are not covered here.[13] Quarter sawing logs, a different method, is discussed later in this chapter.

Being familiar with these procedures will not make you an expert, but should enable you to understand the sawyer's description of how the logs will be sawn, to ask informed questions, and to evaluate the answers. You might want to interview other sawyers if the answers seem vague, or if the person you are considering seems unfamiliar with the whole idea of sawing for grade. By contrast, the sawyer who can incorporate these basics into a description of how your logs will be cut—one who knows from experience both the art and the techniques of sawing for grade—is the one to hire.

One of the most important principles is to saw logs so that as many of the worst defects as possible end up at the edges of the boards, not on their faces. Edge defects can then be trimmed off later leaving clear (or at least clearer) faces which, by commercial standards, means higher quality and more valuable lumber. Surface flaws on the log, such as knotholes or unnatural bulges, are indicators of defects that lie beneath the bark in the sapwood or even in the heartwood. Once the flaws have been located and marked, the log should be rotated into a position on the band mill bed where the first, or opening, cut on top will leave as many defects as possible on the edges of the first board sawn, as illustrated in Figure 53. The opening cut is the most important one because it determines where the remaining three cuts will be made, converting a round log into a cant. The opening cut pretty much fixes the maximum quality of the remaining boards.[14] As illustrated in Figure 54, if it is made haphazardly or without specific regard to the location of defects, all of the lumber sawn from the log will be of lower quality and value than otherwise possible.

If defects are distributed over a wider surface of the log, or even over its entire surface, the area with the worst defects should be turned up on the mill and sawed first. For mills that come equipped with a de-boarder, the log should be positioned with the narrowest end toward the mast and blade. The blade will enter the log somewhere past the narrow end and exit from the broad end. This cut will produce a long triangular slab, several inches thick at the broad end of the log, and tapered to a slightly rounded point at the other end. The de-boarder can easily catch the thick end of the slab and pull it off the log as the mast returns to the starting position for the next cut (if the log position was reversed, the de-boarder is more likely to ride over the top of the slab, since there is no blunt end to catch).

This cut also produces an opposing triangular but flat face on the log that does not extend to the very end. While the face does not have to go from one end to the other, it must extend far enough from the broad end to prevent the log from rocking back and forth lengthwise once it is turned over and the second cut is made on the opposite side.

For the second cut, the log is rotated 180 degrees, so that the first face is lying flat on the mill bed and the best face is up, ready to be sawn parallel to the bed. The log is then rotated 90 degrees. The third poorer side is sawn and then the log is rotated 180 degrees for the fourth and final cut. After the final cut, the sawyer can go back to the first face, and complete what was left incomplete on the first pass; that is, saw a flat face from one end of the cant to the other.

When the best faces are approximately opposite one another, the log should be positioned on the mill so that one good side is up, ready to be sawn, and the other is resting on the mill bed. Unlike the previous cut, the narrow end of the log must be raised so that, lengthwise, the good side up is parallel to the mill bed. Most band mills have either a taper attachment for the mill bed, or a manual or powered lift that will elevate the narrow end of the log (this means that the narrow end should be positioned over the taper attachment or lift, so that it can be elevated). As illustrated in Figure 55, the log is sawn down to where the grade of the last board either drops, or appears as if it is about to drop, down to what would be sawn from the adjacent poorer faces. At this point, the log is rotated 180 degrees so that the second good side is now facing up. The taper attachment or lift is left in position so that this side is also parallel lengthwise to the mill bed.

As shown in Figure 56, successive cuts are made to this side until the grade drops or appears about to drop. Sawing opposite good faces parallel to the mill bed leaves the wedge-shaped cant illustrated in Figure 57. Taper can be removed by sawing several short but successively longer boards from one of the good faces down to the pith, and then rotating the cant and sawing to the pith on the other side. This would leave a thin wedge-shaped board containing the pith, and that can be discarded. Or, to save rotating the cant one last time, it could be sawn down to the mill bed as illustrated in Figure 58. The saw dogs (five vertical posts shown in Figures 55, 56, and 57). on most band mills can be lowered and the board edges can be pressed securely against metal stops protruding from the bed (shown in Figure 58). Either way, the short pieces are bookmatched (the grain on one is the mirror image of the grain on the adjacent piece) and suitable for side-by-side drawer fronts for tables or chests.

The guiding principle in sawing for grade is always to saw the cant in the way that maximizes the quality or grade of the boards. Sawing poor faces, beyond what is necessary to create a stable surface for subsequent sawing, reduces the quality of what the log can otherwise yield. Instead, boards should be sawn from the best face(s) until the quality of what is being cut drops, or appears as if it is about to drop, below what can be sawn from an adjacent side.

Curvature or sweep in the trunks of urban trees is not uncommon. Some is removed when the trunk is bucked into logs. However, enough curvature often remains to warrant sawing the logs into

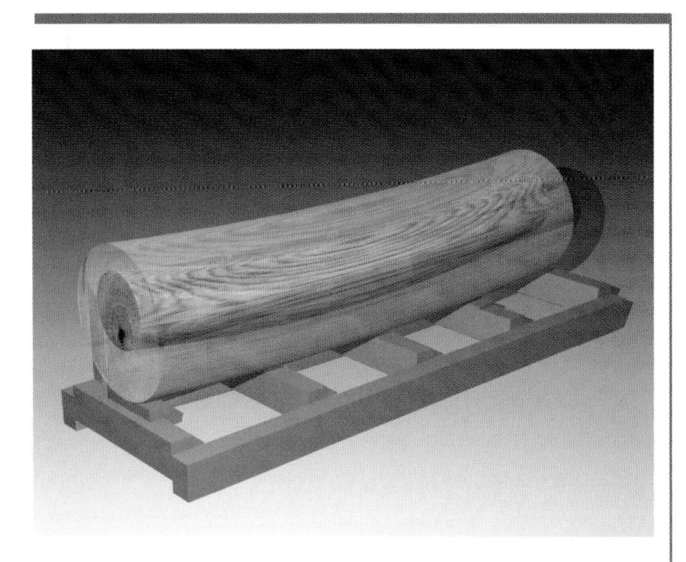

FIGURE 59
In this virtual log, the sapwood is transparent revealing the upward curvature or sweep of the heartwood.
Computer-generated images by Carey Sherrill

FIGURE 60
Curvature or sweep is removed by first sawing aflat face on the side where the log curves upward. This leaves a flat stable face for subsequent cuts. The second cut is made on the opposite side. Some heartwood appears in the middle of the slab since the second cut passes through the belly of the sweep.

FIGURE 61
The first face is then turned toward the dogs and pressed securely against them. Boards are then sawn deeply from the third and fourth faces.

cants in a way that compensates for sweep. The log is placed on the mill with the curved side up and in position to be sawn first, as shown in Figure 59. Since only the middle section of the log is resting on the mill bed, either the taper attachments or the lifts should be positioned under one or both ends of the log, so that it will not rock back or forward while the opening cut is being made. The log should be securely pressed against the saw dogs. As illustrated in Figure 60, sawing curved logs for grade starts with an end-to-end opening cut that goes just deep enough to create a flat stable face. (This produces a slab that is saddle-shaped, thick at each end and thin in the middle.) The log is then rotated 180 degrees, and the second face is sawn from end to end (heartwood will appear in middle of the slab, since we are sawing into the lowest point of the sweep). As shown in Figure 61., the two-sided cant is then rotated 90 degrees so that one of the flat faces can be pressed securely against the saw dogs. The third side is sawed deeply. The cant is rotated 180 degrees, and the sawing completed. Grade is maximized because we get longer boards and better grain than we would from any other cutting method.

Urban trees, especially yard trees, are often felled only after they outwardly appear to be in trouble and are overtly hazardous to nearby houses and yards. The trunks of these trees frequently contain rot but also enough good wood to be worth salvaging. As with surface defects, the opening cut should be nearest the rotted portion of the log and deep enough to create a stable face for the second cut. The procedure then is to saw boards from the better faces until the rotted area is reached. The reduced cant containing the rot can be cut for firewood or ground into mulch.

In general, the quality of lumber sawn from a log depends both on the quality of the log and the skill of the sawyer. Even the best sawyer cannot raise the quality of lumber beyond what the log intrinsically contains. Defects can be minimized but not erased through skillful sawing. On the other hand, unskilled sawing can definitely produce lumber below the quality of what the log contains. Getting lumber of the best quality from urban trees is essential to the whole idea behind harvesting urban timber. Making the very best use of urban trees means not only treating them as something valuable but also getting the best and most valuable wood from each and every tree.

Once you establish the cutting order, and understand how the best boards can be sawn, you must then decide precisely how you want the cants to be cut. This comes down to selecting the exact dimensions. What you decide depends on whether you have something specific in mind for the lumber. In general, you can cut to commercial hardwood and softwood lumber standards, architectural standards, or to individual requirements. Each of these is described below.

Hardwoods: National Hardwood Lumber Association

By commercial standards, green (undried) hardwood thicknesses are measured in quarter-inch increments starting at a nominal 1 inch or four quarters (4/4): five quarters (5/4) is 1-1/4 inches, six quarters (6/4) is

1-1/2 inches, eight quarters (8/4) is 2 inches, and twelve quarters (12/4) is 3 inches. Actual thicknesses are generally one-eighth inch more than the nominal thicknesses. For example, a 4/4 board should be sawn 1-1/8 inch thick to allow for drying shrinkage. Surfaced by a planer on one side, the same board is reduced to 7/8 of an inch. When surfaced on both sides, it is further reduced to 13/16 of an inch. Widths and lengths are described as random but are cut from logs in the way that maximizes the value of the resulting boards (which may entail re-cutting the boards to shorter lengths). Since each log and the boards cut from it are different from the others, widths and lengths vary in a way that seems random.

Prices are quoted per thousand board feet and for thicknesses that run from four to sixteen quarters. Large-scale commercial buyers typically purchase hardwoods by the tractor-trailer load, usually 10,000 board feet. The *Weekly Hardwood Review* and *Hardwood Market Report* provide weekly estimates of hardwood prices by regions of the country (northern, southern, western, and Appalachian). Prices quoted in these two publications are for graded, green, and rough hardwood lumber cut in random lengths and widths.[15]

Graded lumber is classified by quality at the time the boards are cut using standards set by the National Hardwood Lumber Association (NHLA).[16] The underlying assumption of hardwood grading is that the best of the graded boards will be cut into clear rectangular pieces (called clear cuttings) and used in the production of final products, from furniture to flooring to cabinetry. Hardwood boards are graded by the appearance of the poorer of the board's two faces (whereas softwoods are generally graded for strength, since most softwood lumber is used in construction). The standard grades are given in Table 5.

TABLE 5
Hardwood Standard Lumber Grades

Grade	Minimum Width	Standard Lengths
FAS (Firsts And Seconds)	6 inches	8 to 16 feet
F1F (FAS One Face)	6 inches	8 to 16 feet
Selects	4 inches	6 to 16 feet
Common	3 inches	4 to 16 feet
No. 1		
No. 2A		
No. 2B		
No. 3A		
No. 3B		

Lumber down to Number 2A Common is used in furniture, cabinetry, and flooring. Pallets and containers are made from Number 2B (more rot and stain are allowed in 2B than in 2A) and Number 3A Common. Lumber graded 3B is used as loose packing material.

Using the poorer of its two faces, a board is graded by the percent of its surface clear of defects. Among the defects that count are splits, knots, knotholes, wane (either bark remnants or where wood is missing), decay, and holes created by insects and birds. As an example, a board at least 6 inches wide and 8 feet long that is between 83.3 percent and 100 percent clear of defects would be given the highest grade of FAS (first and seconds). A board at least 3 inches wide and 4 feet long, from 66.7 percent to 83.3 percent clear, would be graded Number 1 Common. Neither moisture content (MC) nor surface smoothness are part of the this standard. The buyer and seller, as part of the contract between them, must separately determine both.

NHLA grading rules are precise, very detailed, and are adapted to the characteristics of the species being graded: for example, because of how and where the trees often grow, the rules for grading black walnut are a bit different from those for other hardwoods such as oak, maple and cherry. The Association offers short courses for sawyers (and others) who need to understand grading so that they can saw to grade. However, completion of a full-time 14-week course and several years of field experience are required to become a NHLA-certified grader capable of quickly grading large amounts of lumber, and accurately applying Association rules and definitions to a variety of situations including grade disputes between buyers and sellers. While a grader can grade your lumber to NHLA standards, you cannot claim that the results are NHLA-certified unless the grader is an Association member. Where grading is necessary for a sale, the NHLA can either make a certified grader available or provide the name of one in your area.

Green lumber has not been air dried or dried in a kiln. Rough-cut lumber has not been dressed (planed smooth on one or both board faces): the board faces are rough and saw blade marks are visible. As I pointed out above, widths and lengths are not specific; boards are sawn from logs in a way that yields the most lumber of the highest grades.

Green hardwoods sold widely throughout the country are ash, basswood, beech, birch, cherry, hard and soft maple, red and white oak, poplar, and walnut. Kiln-dried hardwoods are also sold in large quantities for roughly half again their green price per thousand board feet. Kiln-dried walnut may be steamed as well. This means that it has been dried in a way that causes the cream colored sapwood to turn a brown color that approximates the deeper browns of the heartwood. Doing this makes the lumber more uniform in color and more valuable than it would be otherwise. While uniform color may be commercially desirable, to my eye the result is often bland wood that has lost the unique colors and warmth most often found in air-dried walnut.

Softwoods: American Lumber Standard Committee

Hardwoods are used for furniture, cabinetry, flooring, molding, architectural millwork, railroad ties, landscaping and building timbers, pallets, skids, shipping boxes and crates, and cable reels. Some softwoods are used in the same ways, but most are used in construction and, therefore, are graded by a different standard established by the American Lumber Standard Committee. The ALSC operates under the auspices of the National Institute of Standards and Technology of the U.S. Department of Commerce.[17] Though a voluntary standard, it is widely used throughout the United States and Canada. Softwoods sold as construction lumber for structural applications must be graded and stamped by someone who is a member of one of the several regional softwood grading associations (see Appendix D for a list of U.S. associations).

Softwood lumber is either used in construction or is further processed into products ranging from pencils to ladders to millwork and furniture. Construction lumber is graded three ways: for strength only, by appearance alone, and by a combination of the two. Lumber judged only by its strength is visually or mechanically graded for use in construction as joists, studs, and other components of light framing. Design values are criteria for grading lumber at the sawmill. For a particular species such as Douglas fir, or a commercial grouping of species such as southern yellow pine, these values are based on the presence of significant defects—knots, for example—that compromise the strength of the lumber.[18] Softwood lumber is graded and stamped at the mill so that all subsequent buyers down to contractors and carpenters know exactly what they are purchasing.

Through residential building code standards (CABO and IRC™), ALSC design values are incorporated into local building codes that, in turn, determine how graded softwoods can be used in construction. This means wood that you cut in order to construct any structure subject to local inspection and approval will have to be professionally graded before it can be used. Otherwise, local building departments will not approve the plan filed with them before construction begins, or will not approve the structure if it is being built with your own ungraded lumber.

Before using your own lumber for construction, you should check with your city or county building department to see what your local building code dictates regarding the use of ungraded lumber in construction. The department might have inspectors familiar enough with wood and wood grading to judge your lumber. If your wood meets or exceeds the minimum dimensions for framing (for example, your two-by-tens are really 2 inches by 10 inches and not 1-1/2 by 9-1/4), is clear of defects, and has about 20 percent MC, then the inspector might approve it for construction. If the inspectors are not willing, then you might be required to have your lumber graded first by a certified organization. Appendix D is a list of U.S. agencies, accredited by the ALSC's Board of Review, which can provide softwood-grading assistance.

As an alternative, the building department might accept the evaluation of a state-licensed architect or structural engineer. Like most, the building department in my county expects to see lumber specifications on the plan submitted for their approval. If ungraded lumber is to be used, the owner/builder must have the lumber graded for plan approval either by an engineer licensed by the State of Ohio or by a licensed lumber grader. Ungraded lumber used in a building already under construction (presumably substituted for graded lumber specified on the submitted plan) either must be graded before construction continues or it must be removed.

Softwoods are measured and their dimensions expressed in inches and fractions of an inch. For 3/4-inch thick boards, common nominal widths start at 2 inches and run to 12 inches. Lengths start at 6 feet and, in 2-foot increments, go to 24 feet. An 8-foot pine rough-cut into two-by-fours, after being dressed (planed on all four faces of the board) and dried, will yield two-by-fours that are still 8 feet long but will otherwise measure 1-1/2 by 3-1/2 inches. The real thickness of a softwood board is less than the nominal thickness used to identify it.

According to Tom Hanneman, Vice-President and Director of Quality Services of the Western Wood Products Association (WWPA), most softwoods grow in the western and southeastern United States. in forests outside urban areas.[19] In general, when large tracts of softwoods are cleared for development in metropolitan areas, contractors will sell the trees (referred to in the lumber industry as stumpage) to a sawmill. Except for the few trees on small lots, which might end up in landfills, Tom's judgment is that most end up being sawn into lumber or, more likely, used for pulp.

Softwood prices are available from *Random Lengths Lumber* and *Panel Market Report* and price trends and market news from Random Lengths Publications' *Through a Knothole*.[20]

In general, sawing for the commercial hardwood or softwood markets means having to meet industry dimensional, moisture content, and grading standards, and in large quantities. This could be a formidable task for all but the largest urban tree-milling operations. As an alternative, green rough lumber could be sold to a large sawmill or dry-kiln company who would dry and dress it (plane it to size). The company could then combine the smaller quantities with the firm's own output to meet the large quantities demanded by commercial buyers. To allow for shrinkage and dressing to size, green hardwood would be sawn 1/4 inch over nominal sizes: for example, a four-quarter board would actually be cut 1-1/4 inch thick instead of 1 inch thick. Softwoods would be sawn 1/4 inch over their final dry and dressed sizes: hence, a two-by-four would be cut to 1-3/4 inches by 3-3/4 inches so that finished it would measure 1-1/2 by 3-1/2inches.

However, you do not have to stick to commercial quantities and dimensions if you have buyers who want lumber custom sawed to their specifications. Some buyers will use commercial market prices and basic dimensions, but may be willing to buy in quantities smaller than a tractor-trailer load. For example, I found a manufacturer of dowels and spindles in my region willing to buy rough, green, ungraded hardwoods at prices quoted in the *Hardwood Market Report*. One of their employees, a qualified hardwood grader, grades the wood and the company then pays the prevailing price for each of the grades delivered. In a slight modification of basic commercial dimensions, they requested lumber cut 1/4 inch over five quarters to accommodate the final dimensions of their end products.

Hardwoods and Softwoods: Architectural Woodwork Institute

The Architectural Woodwork Institute (AWI) has established voluntary standards for both hardwoods and softwoods used in custom cabinetry, paneling, shelving, ornamental pieces (for example, mantels, fluted pilasters, columns, and corbels), doors, doorframes, windows, blinds and shutters.

The architectural projects themselves are graded as economy, custom, or premium. Economy grade meets the minimum AWI standards for craftsmanship, materials, and installation. Premium meets the highest standards, while custom—the most widely followed of the three—falls between economy and premium.

AWI sets its own lumber grading standards.[21] Three grades, designated as I, II or III, are applied to both hardwood and softwood. Grade I contains the fewest flaws, while grade III contains the most that are allowable. Grade II falls between them. Flaws are those identified by the NHLA (National Hardwood Lumber Association); basically, breaks in the wood such as knots, splitting and checking, or bark pockets, as well as any discoloration. Whether a board of a minimum size (measured in square inches) from a particular species of tree is graded I, II or III depends on the presence of one or more of these flaws. For example, to be classified as Grade I, a walnut board smaller than 275 square inches (10 inches wide by 27.5 inches long) must be flawless. In addition, the board cannot contain more than 5 percent sapwood.

AWI standards and those described above for hardwoods and softwoods are not directly comparable. The AWI standard is concerned only with what a board of a given size looks like when it is put to use in one or more architectural applications. The other appearance grading standards are concerned with how much flawless wood could be cut from that same board.

I have provided a very brief description of NHLA, ALSC, and AWI grading rules to give you an idea of what is involved in selling urban lumber by one or more of these standards.[22] Selling by any one of them will require the assistance of an adviser/grader familiar with the particular standard and with its actual application in customary transactions.

Even though these rules are widely used, they should not be followed slavishly. Buyers are often willing to pay as much if not more for lumber sawn to fit their requirements even when the dimensions deviate from the rules. In addition, sawing to the buyer's needs reduces the amount of trim waste that arises from sawing rule-based dimensioned lumber.[23]

Cutting to Individual Requirements

When sawing for myself, I try to have a project in mind, so that the lumber can be rough-cut to match the project dimensions. While sawing, I keep track of boards that have particularly interesting grain that, if carefully arranged, might make an interesting tabletop, matching cabinet doors, or drawer fronts. If the grain is especially appealing, I cut thin slices off of the board to use as veneer. When it is finely adjusted and equipped with a new blade, I have been able to cut slices down to 1/4 inch thick on my sawmill. For one project, my son and I cut the veneer into eight equal-sized wedges to create the round sunburst tabletop shown in Figure 62 below. Using a drum sander, I smoothed and further reduced the wedges to a uniform thickness of just under 1/8 inch and then carefully glued each wedge to a plywood substrate.[24] We cut the round frame for the top in such a way that the grain of each piece of the frame matched the grain of the adjacent pieces.

If I'm not working on a specific project but have to saw the log anyway, I cut basic parts for the kind of projects I tend to do (usually tables). Watching the grain as the boards come off of the mill, I mark those pieces that will best serve the projects I do. I'm especially interested in grain patterns that will create unique tabletops. For appearance, as well as for dimensional stability, I have most of every oak log I acquire quarter sawn (described below).

When cutting for those who do not know what they may eventually want, I recommend cutting in a way that combines appearance and basic dimensions. Appearance judgments are subjective and are made as the log is cut. I cut 5/4 thickness for tabletops and several 12/4 pieces for table legs (so that I do not have to glue thinner pieces together later on). I do not recommend lengths over 12 feet unless the owner wants a large conference or dining room table.

All of my discussion so far would have us sawing lumber to prescribed dimensions, either to meet architectural or commercial standards, or to match the

FIGURE 62
Round table top made with thin slices of bur oak matched in a sunburst pattern.
Photo by Sam Sherrill

requirements of a specific project or likely projects in the future. However, with fewer words but with equal emphasis, I recommend that you allow the wood to guide what you might be willing to make, as opposed to letting either commercial standards or the project dictate how the log is cut. As the log is being sawn to prescribed dimensions, watch for boards with unusual color, grain, or even what would otherwise be considered a serious defect. Let what you see suggest something unique. Instead of forcing the wood to completely conform to the project, let the project emerge from the interplay of the wood and your imagination. As a modest example, I made the Arts and Crafts style table in Figure 64 specifically for a piece of ash that, as it came off of the mill, was almost thrown away as scrap because it was narrow, split at both ends, and slightly cupped. After trimming the split ends and repeatedly passing the board through my drum sander, I ended up with the unusual circular grain shown in Figure 64 and 65.[25] I built a sofa table to accommodate the board because sofa tables are narrow and long—proportions that just suited this board.

Some logs will yield few, if any, furniture-quality boards. However, even this wood, if sound, can be put to uses other than mulch or firewood. There are a variety of outdoor applications for ties, timbers, planks, and boards cut from the heartwood of lower quality logs (because it lacks the extractives found in heartwood, sapwood in all species is less durable). Landscaping ties, borders for playground areas, fencing, hillside steps and railings for backyards and parks, compost bins, parking lot blocks, and plank decking for trailers are a few examples. Because they are very resistant to decay, black locust, mulberry, and osage orange are well suited to on-ground and in-ground uses. Cypress, cedar, redwood, the various white oaks, and even walnut and cherry are also resist-

FIGURE 63.
The bench above was made by sawing the log down the middle, and through the crotch where two stems meet. The bark side was left round and the two supports were cut to match the curvature of the log. This simple construction produces visually appealing garden seating.
Photo by Patricia Sherrill

ant (see Appendix E for a list of domestic woods categorized by heartwood resistance). One indoor use I found was at a local stone-cutting business. Large blocks of stone are set on timbers than in turn sit like joists on very large concrete supports. The saw blade is set to cut through and just below the stone and into the wood timbers. Cutting the timbers prevents damage to the blade and to the permanently-positioned concrete supports. There is a constant demand for new timbers as the ones in use are ultimately cut so many times that they must be replaced.

Another potentially significant outlet for lower quality but sound wood (mainly hardwoods) are firms that manufacture and repair pallets, bins, shipping crates and boxes, and cable reels. The pallet and container industry uses a significant amount of the annual output of American hardwood lumber. In 1999, the industry used 4.4 billion board feet of hardwood or about 32 percent of the 13.8 billion board feet consumed. The industry used 2.1 billion board feet of softwood or only about 4 percent of the 54.5 billion board feet consumed.[26]

FIGURE 64.
A sofa table was made of ash and oak to feature the unusual ash board used on the top.
Photo Sam Sherrill

FIGURE 65.
The light reflected from the middle of the ash top rotates as the observer moves from one end of the table to the other. This unusual quality is referred to as chatoyance, a shifting pattern of light usually found only in polished gemstones described as cats' eyes.
Photo by Sam Sherrill

The National Wooden Pallet and Container Association can provide the names of companies in your state and county that manufacturer or recycle these products.[27] Contact the company before cutting to see if they would be interested and, if so, what dimensions and species they are willing to purchase and, of course, what they are willing to pay.

Whatever the ultimate uses, a log can be sawn three basic ways: in cross-section, tangentially, or radially. See Figure 66. Bucking a tree trunk into separate logs is cross-sectional cutting. Cutting the length of a log one slice after another entails cutting on the tangent to the circular growth rings. Radial sawing means cutting boards along the radius of the log across the growth rings. For a given species, the way in which a board has been sawn influences its surface appearance, how it dries, and later (in a piece of furniture) how it looks and how it responds to changes in humidity. The growth rings at each end of a board, referred to as end-grain, reveal the way it was sawn from the log.

Flat Sawing

Tangential sawing from the outside of the cant straight through from one end to the other initially yields what is commonly called flat sawn lumber (sometimes referred to as plain sawn). As illustrated in Figure 67, the growth rings in such a board form concentric semi-circles around the pith or center of the log. Cutting tangentially to the tree's growth rings creates, on the faces of a board, what is commonly described as a cathedral pattern.

The cross-section of a flat sawn pine board in Figure 69 shows the semi-circular end-grain of the board. You can see the cathedral pattern in the ash board at the top of Figure 68.

An advantage of flat sawing is that it produces the widest possible boards that can be cut from the cant, the very widest

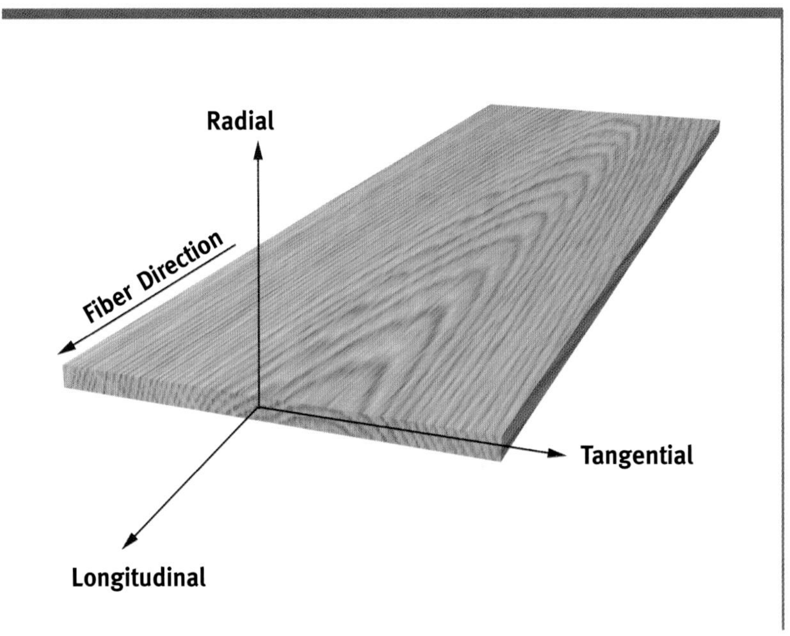

FIGURE 66.
The three axes of a flat sawn board: radial (through the thickness of the board); longitudinal (its length); and, tangential (across its face).
Computer-generated image by Carey Sherril

being cut right through the center (which would produce a board that is mostly quarter sawn, described below, except for the very middle). Another is that the cathedral pattern is visually pleasing and works well when two or more bookmatched boards are used in visible places such as tabletops.

An important disadvantage of flat sawing is that, as the boards dry, they tend to cup across the board face, away from the center and toward the bark side. This is why, as a rule when flat sawn lumber is used as decking, the boards are fastened to the joists bark-side down. The nails or screws used to fasten the board to the joist will minimize cupping as the board loses moisture. Fasteners reduce cupping but cannot completely prevent it, especially when the pith of the log is at or near the center of the board. I alternate flat sawn boards up and down when gluing them together for a tabletop: that is, one board is bark side up and the next glued to it is bark side down, then the next one is up, and so on. The theory behind alternating boards up and down is that slight cupping in one direction will be offset by the slight cupping, in the other direction, of the adjacent board. In my experience, this works well most but not all of the time. Narrower boards further reduce cupping but also interfere with the side-to-side continuity of the cathedral grain pattern.

I avoid boards and posts that contain the pith at the center of the log because I know they will cup in one direction or another, very often to the point of splitting. The posts will twist and no fasteners I have ever used will prevent this from happening. In my view, this is just not usable lumber. When I cut a cant, I "box the heart;" that is, I alternately cut and rotate the cant down to a single square post that contains the pith. I also rip wide

FIGURE 67
Growth rings form semicircles at the end of flat sawn boards and a cathedral pattern on the board face.
Computer-generated image by Carey Sherrill

flat-sawn boards into two or more pieces, especially if they were cut near the pith. After the boards have dried, I glue them back together. I also saw the boards at least 3/8 inch thicker than I ultimately need so that I end up with a flat board at the correct thickness. I run the cupped board through my thickness planer with the cupped (bark) side down first, flattening the opposite (pith) side. Feeding it into the planer the opposite way—cupped side up—may produce a board of uneven thickness, because the board will rock from side to side on its curved face as it passes under the blades. Once the pith side is flattened, I turn the board over and flatten the bark side. If the board has cupped from center to edge by 1/8 inch then, after planing each side by this amount, I'm left with 1/8 inch for final planing and sanding.[28] Cutting this way does turn some potentially useful lumber into sawdust.

Quarter Sawing

As illustrated in Figure 70 quarter sawing, in effect, entails cutting each board along the radius from the outside to the center of

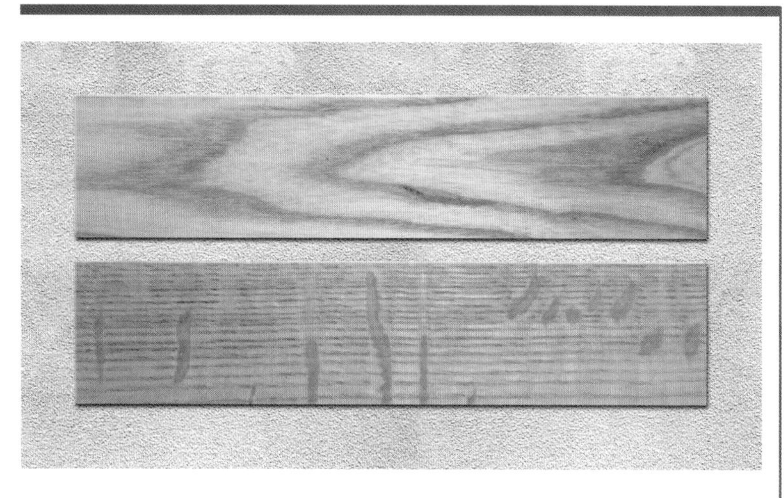

FIGURE 68
Flat sawn ash board on top and quarter sawn red oak on the bottom. Notice the short, irregularly spaced, and smooth ribbons running across the long grain of the oak board. This is referred to as ray flake.
Photo by Sam Sherrill

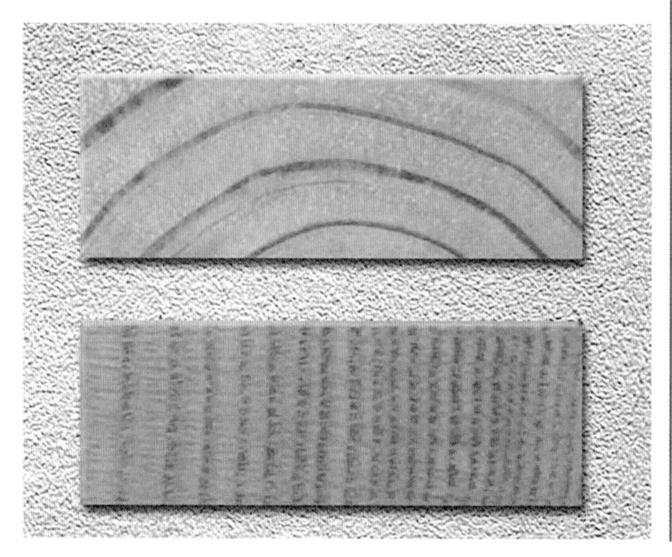

FIGURE 69
Cross-section of a flat sawn pine on top and of quarter-sawn red oak on the bottom. Notice wide, light-colored earlywood in the pine.
Photo by Sam Sherrill

the log (as a general rule, this cut works best for logs 18 inches in diameter and greater at the smallest end). Cutting radially across the tree's growth rings yields boards with nearly straight growth rings approximately perpendicular to the board faces as shown at the bottom of Figure 69. Quarter sawing creates the linear pattern of roughly parallel earlywood and latewood on the faces of each board shown at the bottom of Figure 68. In oak and sycamore, this method of sawing produces visible ribbons of smooth grain called ray fleck (the result of cutting parallel to the ray cells that run from the center of the tree to the sapwood). Ray fleck is visible in the oak board in Figure 68.

Instead, the log can be cut into quarters—hence quarter sawn—and then each quarter section sawn across the growth rings. Even this would require an awkward positioning and clamping of each quarter section. There is an easier way to do this. As shown in Figure 71, cut the log into three sections, A, B and C. Push the top section, A, off to the side and cut section B into boards. Then set these boards aside and, as shown in Figure 72, rotate section C up 90 degrees so that the long flat side rests against the log stops on the left side of the mill. Cut C into boards.

The boards can then be re-stacked and edge-trimmed together or they can be individually cut later. (I usually cut each board later, thus getting slightly more width than if I cut them all at once to the width of the narrowest board.) Next, as shown in Figure 73, place section A back on the mill and cut it the way section C was cut. Stack these boards with those from C. Finally, as shown in Figure 74, place the boards cut from section B back on the mill up on end and cut them slightly above and then below the pith. These two cuts box the

FIGURE 70
Growth rings are nearly vertical to the faces of these quarter-sawn boards. On the board face, the growth rings form nearly straight and parallel lines of alternating light colored earlywood and dark latewood. This image illustrates the meaning of quarter sawing. No band saw mill could actually saw a log this way.
Computer-generated image by Carey Sherrill

FIGURE 71
The log is cut into three sections, A, B, and C. A is set aside and B is sawn into boards.
Computer-generated images by Carey Sherrill

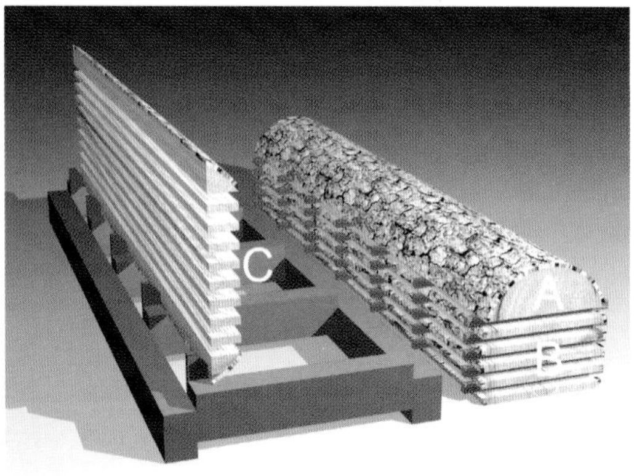

FIGURE 72
Boards cut from B are set aside. C is rotated up 90° and cut into boards.

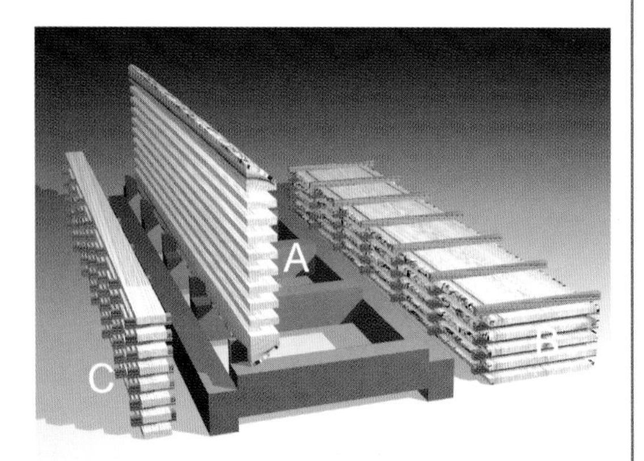

FIGURE 73
Boards from C are stacked. Section A is then cut.

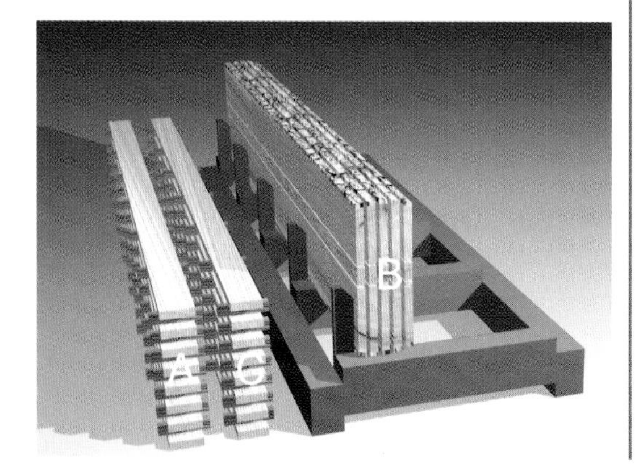

FIGURE 74
Boards cut from section A are stacked with boards from C. Boards from B are turned on edge and sawn twice down the middle. Cutting just above and below the pith boxes the heart of the log.

heart of the log in a single post. Since the boards cut from B are all about the same width, they can be edge-trimmed together without much loss of valuable wood. Close to 70 percent of the log can be quarter sawn this way.[29]

The linear grain pattern in all woods and, in oak and sycamore the addition of ray flake, are also visually pleasing. Quarter sawn oak (fumed with ammonia to produce a dark finish) is one of the distinguishing features of American Arts and Crafts furniture, especially the work of Gustav Stickley.[30] Though it dries more slowly compared to flat sawn lumber, quarter sawn boards are much less likely to cup, split, and otherwise contort themselves as they dry (this is especially true of sycamore). And, with changes in humidity, they expand and contract much less than flat sawn boards.

The only disadvantage of quarter sawing is that the boards are narrower than flat sawn ones. However, I find in my own woodworking that this is not much of a disadvantage and, in fact, is really more of an advantage. Seamless wide tabletops can be made from several otherwise narrow quarter sawn boards. If I carefully match the linear grain patterns of adjacent boards, and there are no visible gaps between them, I can glue up many single boards to make a wide tabletop that looks as if it had been cut as one piece right from the log. Moreover, unlike its glued-up flat-sawn counterpart, I do not have to worry about cupping. In my judgment, there are only two slight disadvantages to quarter sawing: you might forgo an interesting cathedral pattern, and the sawing is a bit more involved than the method that produces flat sawn lumber.

Rift Sawing

The Architectural Woodwork Institute (AWI) considers a board quarter sawn if the growth rings run from 60 to 90 degrees to the face of the board. Flat sawn is zero to 30 degrees to the board face—basically the circular end-grain described above.

Between these two is rift sawing, where the growth rings are between 30 and 60 degrees to the board face. Rift cuts produce linear grain on all four sides of squared boards—say three-by-threes or four-by-fours—that work well for table legs. While the grain is linear, there is no ray flake.

Flat and quarter sawing yields linear grain on two sides, and plain on the other two sides. When stained, the appearance of the two sets of sides is not noticeably different. Whether this difference is displeasing is in the eye of the beholder (this beholder does not like it). Rift cut squares are easy to identify: the end-grain runs diagonally from one corner to the opposite corner. You can see the diagonal grain at the end of the protruding square board in Figure 75.

Others are less specific than the AWI about differences between flat, rift, and quarter-sawn lumber. In general, commercially traded lumber is considered quarter sawn if the growth rings run between 45 and 90 degrees to the face of the board. Anything below 45 degrees is considered flat sawn. For hardwoods, some dealers label lumber between 30 and 60 percent as bastard sawn. Rift sawn is not separately identified or defined.[31] Clearly, if you plan to sell urban lumber, you need to know in advance exactly what prospective buyers mean when they use such terms as flat, rift, bastard, and quarter sawn.

Drying Lumber

Once the logs have been cut into lumber the next step is to dry the lumber. This can be done in a kiln, or by air drying. Each of these methods has its advantages and disadvantages, which I will discuss in this section. But before we get to that, we need to review how the lumber holds and gives up water. This knowledge sheds light on the drying process, how drying alters board shape and dimensions, and how the lumber can be damaged if drying moves too quickly.

Water here means the water content of tree sap along with water that comes from other sources such as the atmosphere or even from being submerged in a river or lake. Lumber freshly cut from a living tree (or one recently deceased but not dried out) is referred to as green wood. Green wood has about the same meaning as the wet lumber referred to above in the discussion of commercial lumber prices. As explained below, there is a specific level of water in wood that separates green from drying lumber.

A moisture meter is essential in measuring moisture content (MC) and keeping track of drying inside a range of about 6 percent to 30 percent (they are inaccurate outside this range). I use an electronic meter that measures to a depth of 3/4 inch. Used from both sides, I can measure a board up to 1-1/2 inches in thickness. I consider this to be one of the most important wood-drying and woodworking tools I own. I never start any project without first measuring the MC of the wood I'm about to use. Whether or not I use it depends on what the meter shows.

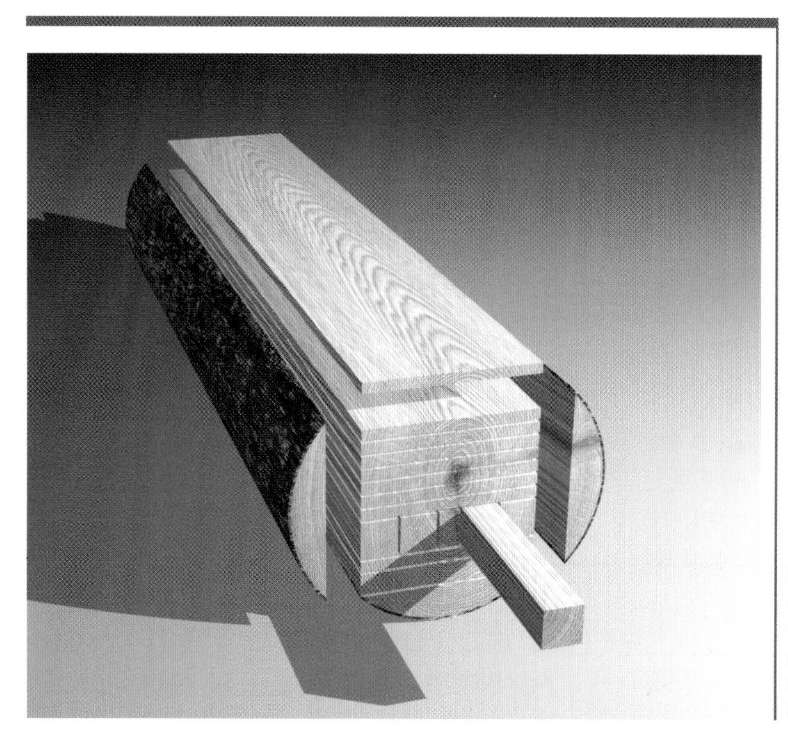

FIGURE 75
Rift-sawn board suitable for table legs, with linear grain on four sides.
Computer-generated image by Carey Sherrill

Water in Wood

While the shapes, sizes, functions, and positions of cells vary throughout a tree, their walls create an enclosed cavity that holds sap (mostly water), air, or living protoplasm. Water is considered bound when it is in the cell walls and free in the cell lumen—the cavity formed by the cell walls. Water molecules in the walls are held between long chains of cellulose molecules. As it dries, the cell first loses its free water from the lumen and then the bound water from its walls. A wood cell is considered to be at its fiber saturation point (FSP) once the free water is gone and only the bound water remains.

The MC of wood is calculated by first subtracting the oven-dry weight of a board (where both free and bound water have been removed) from the weight of the board before drying. Suppose an oak board from a freshly cut tree weighs 20 pounds wet and 15 pounds after being completely oven-dried.[32] The difference between the wet and dry weights is 5 pounds. Expressed as a percentage of its dry weight, the MC of the wet board is (was) 33 percent (20 - 15 = 5: 5 ÷ 15 = .33 x 100 = 33%). Another freshly cut board that weighs 50 pounds wet and 20 pounds totally dry has a MC of 150 percent. In living trees, MC ranges from 30 percent to more than 200 percent. The MC of the sapwood in softwoods is greater than in the heartwood. In hardwoods, the MC of heartwood and sapwood varies by species.[33] For example, for northern red oak, the MC for the sapwood is 80 percent, and 69 percent for the heartwood. By contrast, for sweetgum, the MC is 79 percent for sapwood and 137 percent for heartwood.[34]

To be somewhat more precise about fiber saturation, we can say that lumber has reached its FSP when its MC is between 25 and 30 percent. Lumber above its FSP is considered green, while below it is considered air-dried (even though this is an imprecise term and can even refer to lumber close to 30 percent).

As water molecules begin exiting the cell walls, MC falls below the FSP. Below this point, the cellulose fibers move closer together, filling the space vacated by the water molecules. The cell walls shrink as the fibers move closer together. As the walls shrink, the entire wood cell itself becomes narrower as well. However, the length of the cell remains unchanged (except where they are distorted by compression or tension).

Most tree cells are elongated and are aligned in an overlapping fashion along their lengths, parallel to the long axis of the tree trunk or limb. Most of these cells are vertically oriented along the tree trunk; that is, the long axis of the cells runs parallel to the long axis of the trunk. Others, called ray cells, run at a right angle to the long axis of the trunk, thus are perpendicular to the vertically oriented cells. The ray cells run from the center or pith of the tree to the bark.

A flat sawn board is cut tangentially to the vertical cells and across the ray cells. Thus, this board consists mainly of cells, side-by-side, oriented along the length of the board (just as they were oriented along the length of the tree trunk). As water escapes the cell walls, and the walls narrow, all the cells across the width of the board also become narrower. This explains why flat sawn boards shrink most across their width. Because the cells shrink very little in length, board length remains nearly unchanged (lengthwise shrinkage is about 0.1 percent).[35]

A quarter-sawn board is cut across the vertical cells and tangentially to the ray cells—just the opposite of the flat sawn board. As shown in Figure 68 (the bottom quarter-sawn oak board), the ray cells cross the board faces. Since cells do not lose length as they give up their bound water, the width of a quarter sawn board narrows less than a flat sawn one as it dries. A quarter sawn board is flat sawn on its edges and can, therefore, be expected to shrink more in thickness than a flat sawn board during drying. As shown below in Figure 76, shrinkage in quarter sawn board A is uniform compared to the more pronounced cupping of other cuts such as C and B. While quarter-sawn boards dry at a slower pace than flat sawn, they are much more likely to retain their original shape.

As you can see in Figure 76, as boards are cut from the vicinity of C toward the center of the log, the end-grain of the boards becomes increasingly perpendicular to their faces. At A, the grain is perpendicular; thus the board is quarter sawn. The reason a board like C can cup toward the bark side of the tree is that it is more flat sawn on the bark side and more quarter sawn on the center side. The bark side shrinks more than the center side, thus pulling the edges of the board outward. For a given thickness, cupping becomes more likely and more pronounced in boards cut closer to the center. This tendency is even greater in logs of lesser diameter cut from younger trees. The reason wide boards sawn from larger logs, cut from correspondingly older trees, do not cup is that for a given thickness there is less difference between the bark and center side shrinkage. Wide boards from old-growth trees used in furniture made several centuries ago are still flat.[36]

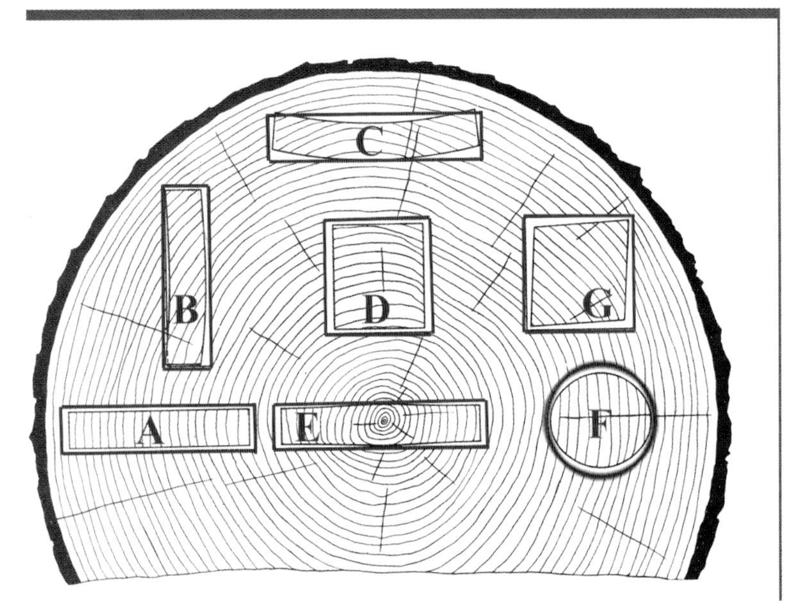

FIGURE 76.
In cross-section, A through F are the cuts that can be made from a log. Viewed from the end, the red outline is the shape of the board just after it is cut. The black outline shows the distortion that can occur after the board has dried. The round shape, F, is of course not cut by a saw. It shows how a spindle or dowel would flatten from the top and bottom as it dries. G is a rift-sawn square that has taken on a slight diamond shape. Even though it is mainly quarter sawn, E contains the heart and is likely to warp and split right through the pith at the center of the heart.
(Forest Products Laboratory, USDA Forest Service.1999. Wood Handbook, Wood as an Engineering Material, General Technical Report, FPL-GTR-113, U.S. Department of Agriculture, Forest Service, Forest Products Laboratory, Madison, WI, p.3-8)

Figure 77 illustrates the other ways boards can become misshapen during drying.

A board does not dry evenly throughout its thickness. It dries from the outside to the inside and from the ends toward the middle. Water escapes more rapidly through the ends of the boards, where the vertical cells have been cut, than it does from the faces of the boards. Water also evaporates from the surfaces of the board, pulling moisture from its wetter center. The difference between the amount of water in the interior of a board (the core) and its surfaces (the shell) is referred to as a moisture gradient. If the gradient is slight—if there is little difference between the surface and interior moisture content—then the board will dry very slowly. If the gradient is pronounced—if the difference is substantial—then the board will dry more rapidly. If the gradient is too great—if the difference is excessive—the board could come under enough stress to cause irreparable damage that shows up as cracking on the surface and ultimately throughout the board. Once a board is riddled with cracks, it has lost both strength and appearance and is useful only as firewood. Controlling the pace of drying is very important: too slow and we do not get to use it soon enough, too fast and we do not get to use it the way we want.

Why dry wood? As it drops below the FSP, wood becomes lighter and shrinks dimensionally. Yet, at the same time, it gains strength and becomes more resistant to decay, insect, and fungal damage. Dried wood is less expensive to ship than heavier wet wood and, once it achieves a stable MC (and fixed weight and dimensions), it is ready for all the uses we put it to.

Once a stable MC is achieved, does it remain fixed? Wood is hygroscopic: this means that it readily exchanges water with the air around it (it absorbs and desorbs

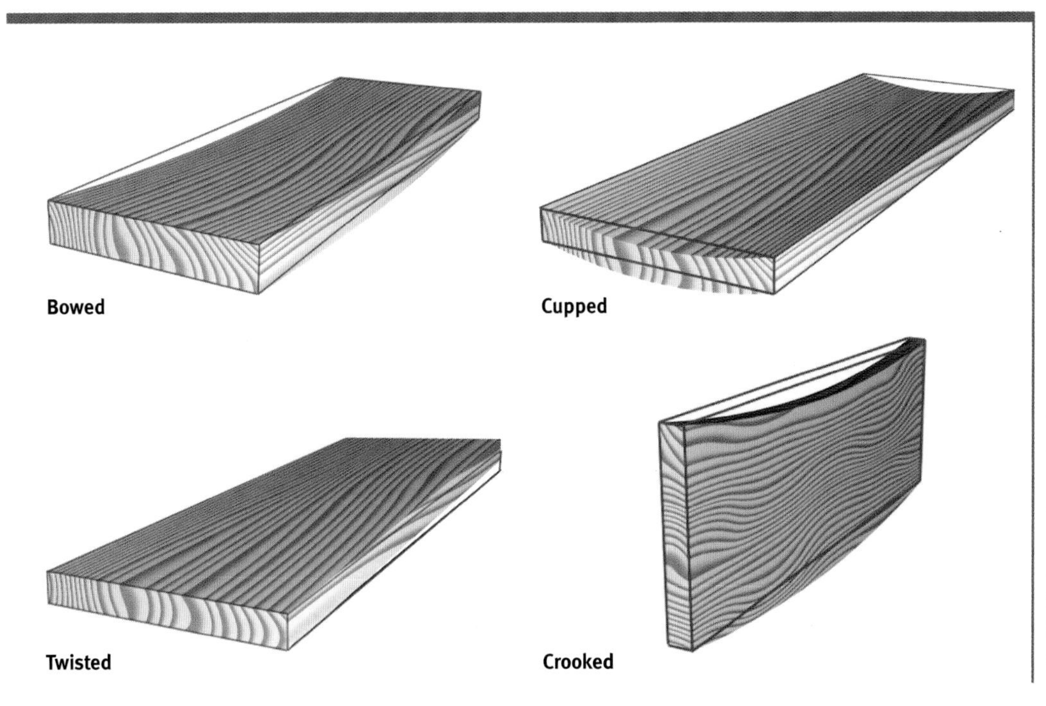

FIGURE 77. When good wood goes bad. Four common ways wood can distort itself during drying. **Computer-generated image by Carey Sherrill**

water). The direction of this exchange—whether the wood taking in or giving up water—depends on the MC of the wood and the relative humidity of the surrounding air.[37] As long as relative humidity is constant, wood will eventually dry down to an equilibrium moisture content (EMC).

The lumber I have stored in my basement workshop loses bound water until its MC is in equilibrium with the relative humidity of my shop. The lumber MC remains fixed as long as the relative humidity of my shop does not change. So, one answer to the question is yes: MC remains fixed as long as the relative humidity of the surrounding air remains constant. In the winter, when I run my gas furnace, the relative humidity in my basement drops and so does the MC of my lumber. In the summer, even though my home's central air conditioner keeps the temperature and the relative humidity in my house lower than they are outside, the humidity in my shop is still higher than it was the previous winter. When the relative humidity of my shop rises, so does the EMC of my lumber (again, in the form of bound water). Typical wood finishes, such as several coats of polyurethane, can appreciably slow the exchange of moisture between wood and air but not completely stop it (oil and wax finishes have no effect on the exchange).[38]

That the wooden screen door sticks in the summer and not in the winter demonstrates how changes in relative humidity from one season to the next alters not only the MC of the door but its dimensions as well. When things made from wood absorb water from the air, they expand, and when they give up water, they contract. The door expands just enough in the summer to stick in its frame, and contracts just enough in the winter to open and close smoothly. (I have noticed that these doors are always painted on both faces but rarely on the top or bottom, thus assuring that they will expand and contract.)

One summer I built a sofa table, using as the top a single piece of flat sawn oak that I firmly (but mistakenly) glued to the table frame. One day the following winter, I heard what sounded like a pistol shot in my living room. After poking around for the source of the noise, I finally found the tabletop sitting loosely on its frame. Shrinkage of the top had caused the glue to fail so suddenly and completely that it popped when it let go.

In another example, a piece of furniture made in a humid climate contained what were supposed to be floating panels in a frame at each end (the panels float in the sense that they sit unfastened in their frames). The maker nailed the panels to the frame, an arrangement that would have worked fine if the piece had stayed in its native country. However, in the much lower indoor relative humidity of a cold American winter, the panels shrank in from the sides so much that the wood split around the nails. The unfinished panel edges, hidden in the more humid climate, were revealed as well.

A store near where I live sells unfinished wooden furniture from the British Isles. Made and used in a moist climate, the tops of these tables split open during our cold dry winters. Oddly enough, these substantial fissures in the tops of tables seem to add to their value. I can only surmise that, for the buyers, these drying defects lend authenticity to the piece. The single lesson of these examples is that we must pay close attention to wood as it dries and then to the way we fasten pieces together in the projects we make. Wood will move in response to shifts in relative humidity and we must accommodate, not ignore or resist, that movement both in drying and building.

Figure 77 graphically portrays the direct nonlinear relationship between changes in relative humidity, measured on the horizontal axis, and the EMC (equilibrium moisture content) of wood measured on the vertical axis. The wood represented by the single black line is white spruce: its EMC response to changes in relative humidity is treated as representative of most types of wood.

Kiln Drying

Water can be removed from wood by a dry kiln.[39] In general, a wood-drying kiln consists of one or more closed, thermally-insulated containers or chambers where there is precise control of the interior temperature, relative humidity, and air movement. As the temperature of air rises, so does its moisture-carrying capacity (see footnote 37 to this chapter for a more detailed explanation). In the closed chamber(s) of commercial kilns, temperatures are raised well above the usual ambient outside temperature. Higher temperatures raise the moisture-carrying capacity of the inside air. The more moisture the air can hold, the faster water moves through and out of the wood being dried inside the kiln. For even drying of all boards, air must be evenly circulated by fans among all the boards being dried. The circulating air carries the heat to and also transports moisture from the boards. At some point, the moisture-laden air is vented out of the kiln and is replaced with fresh air, or the moisture is removed from the air and the air is recirculated.

One of several ways to classify kilns is by their operating temperature ranges.[40] Low temperature kilns operate from 70 to 120 degrees Fahrenheit; conventional kilns from 110 to 180 degrees F; elevated temperature kilns from 110 to 211 degrees F, and high temperature ones from 230 to 280 degrees F. Hardwoods are dried mainly in low-temperature and conventional-temperature kilns while some softwoods are dried in conventional and elevated temperature kilns. High temperature kilns are used mainly to quickly dry construction grade softwood lumber, taking no more than a day or two to do so.

Lumber is loaded in stickered stacks in a kiln, a process called charging the kiln. Stickers are strips of wood uniformly placed to create air-circulation spaces between the layers of lumber in a stack. For an example of a small kiln being charged with stickered lumber, go back to Figure 12 in Chapter 1 (the kiln pictured is owned and operated by the California Department of Forestry and Fire Protection). In large kilns, stickered lumber is stacked on kiln trucks and rolled on tracks like railroad cars into the kiln.

As a rule, kilns are fully charged with lumber that, by dimension and species, will dry at about the same rate. Doing otherwise leads to wood being either damaged by the stresses of over drying or not being dried down to the desired MC.

Once charged, kilns follow a drying schedule that in large commercial operations is automatically controlled. The schedule for a particular species of given dimensions would start at a lower temperature and rise at a rate and to a temperature that the boards can tolerate without structural damage. Recall that the moisture gradient is the difference between the amount of water in the core of a board and in its shell. If this gradient is too great—if the temperature in the kiln rises too quickly or too high—the drying stress can cause irreparable damage that shows up as cracking on the surface and ultimately throughout the entire thickness of the boards. Stress can arise when the shell of a board is drying too fast for its wetter core. Cracks begin appearing on board surfaces as the shell tries to shrink more than the core will allow. Kiln operators relieve this stress by wetting the board surfaces with water or steam. The fibers swell as they absorb water, thus relieving some of the pressure on the core. This illustrates the need for careful monitoring of kiln drying even though it is an automated process.

Sometimes lumber is dried down to its FSP (fiber saturation point) before final drying in a kiln. This can be done by air drying in sheds or in warehouse pre-dryers, low temperature kilns that reduce MC to about 25 percent. Either way, pre-drying reduces the time and energy needed for final kiln drying.

For thousands of years, air-dried wood

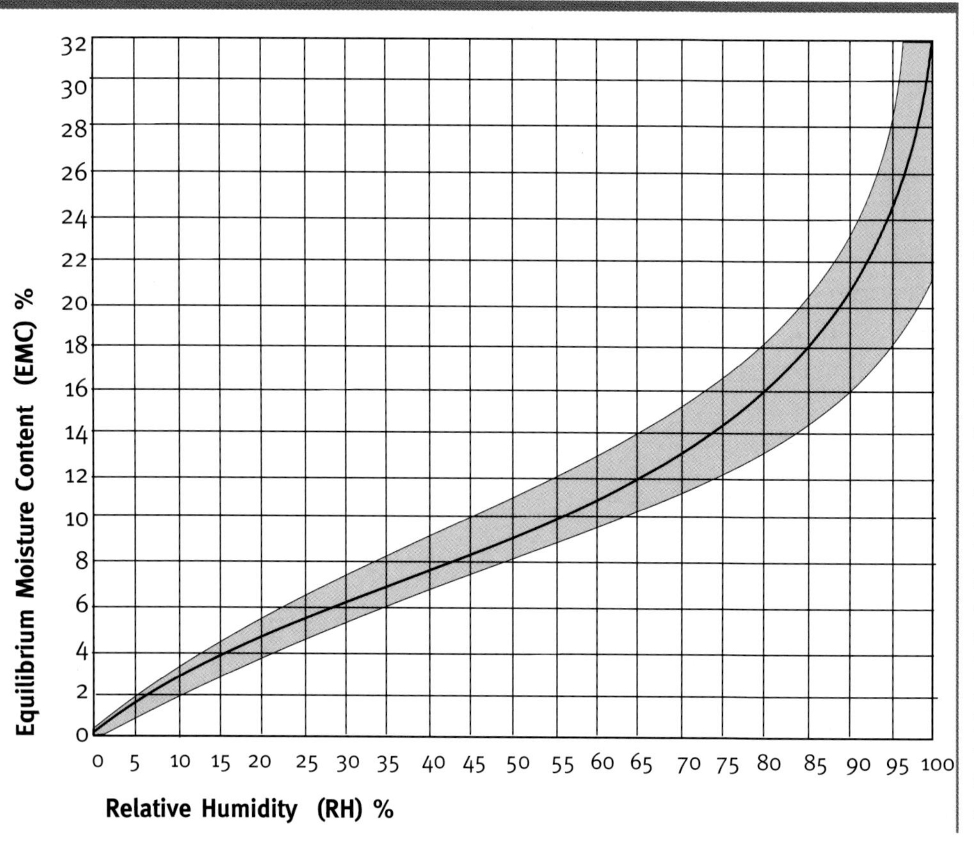

FIGURE 78
This chart graphically represents how the EMC of wood (white spruce in this case) is directly dependent on relative humidity: changes in humidity are followed by changes in EMC. For example, at fifty percent relative humidity, we can expect most woods to dry down to about nine percent. If the relative humidity then fell from fifty to twenty-five percent, EMC would drop to about five percent. By contrast, if humidity jumped from fifty to seventy-five percent, EMC would rise to fourteen percent.
Used by permission of R. Bruce Hoadley

was used as construction material, for flooring, and to make crafts, cabinets, and furniture. So, after all that time, why is it now being dried in kilns? While kiln drying can yield lumber with fewer defects than air drying, the answer has much more to do with economics than the virtues of dryer wood.

Unlike it was in earlier times, lumber is now a standardized commodity that is bought and sold in large quantities. Many buyers never see what they purchase. Instead, a buyer relies on the seller to provide large quantities of lumber that meets the dimensions and standards set in the terms of the sale. One of the standards is moisture content, an important determinant of the strength, stability and appearance of lumber used in everything from construction to large-scale furniture manufacturing. Kiln drying not only meets specified moisture content but also substantially shortens the amount of time needed to convert timber into manufactured wood that has been delivered and paid for. While the heat from kiln drying lowers the MC of wood and reduces the likelihood of insect or fungal infestation, it also gets the wood to the buyer much faster than air drying. Because kilns substantially cut drying time, suppliers can respond to changes in the level of demand for lumber much more quickly than they could if they relied on air drying alone. In addition, suppliers are not forced to invest in large inventories of lumber in various drying stages, as well as the space and drying sheds that would be required. Capital and operating expenses of kilns are not small, but these facilities pay for themselves by minimizing inventories and by their capability to quickly process large quantities of lumber that meet the requirements of the lumber market. Conceivably, a pine tree could be felled, limbed and bucked one day, sawn into two-by-fours and graded the next day, dried the third day in a high-temperature kiln, loaded on a tractor-trailer the fourth day, delivered to a home center or lumber yard on the fifth day and sold that same day. Comparable air drying could take months.

If you choose kiln drying, check Appendix F for the names of kiln-drying associations that can identify a kiln in your vicinity willing to dry at or below truckload quantities (10,000 board feet). There are four important points to keep in mind if this is your choice.

The first and most obvious is that you will have to pay for it. Rates vary by moisture content, species, board thickness, and the amount to be dried (see Appendix G for an example of prices for hardwoods in Ohio). Separate pre-drying might be required if your lumber is above the FSP of 25 to 30 percent. Species are grouped together by their drying compatibility: for example, as a group, ash, cherry, and hard or soft maple can be dried together, as can elm, beech, oak, sycamore and walnut. However, these two groups cannot be mixed. And, the price probably will be higher for the elm group than for the ash one. For a given grouping and thickness, you can expect to pay much lower rates for larger quantities. You will have to sort the lumber into separate stacks by length (for example, 6 to 8 feet, 9 to 10 feet, and 11 to 12 feet). Very likely, there will be fees for unloading and loading, stickering, and sorting (if you have not done these things).

Taking all the pricing factors into consideration, the total cost could be below $200 per thousand board feet, or more than twice as much.

The second point is that you will have to deliver the wet lumber and pick it up after it has been dried. You will need a truck with sufficient load capacity and also some help.

Third, before loading, you should inspect the dried lumber for cracks, especially those on board faces that penetrate deeply into the boards. This could be a sign that the wood dried to quickly or at too high a temperature, or both. If most or all of the boards appear deeply cracked, the lumber could be a total loss, good only for firewood. As you would with any other merchant, ask for references or names of customers and then call and ask for their comments.

Fourth, you will need a place to store the dried lumber, preferably not outside where the wood, being hygroscopic, will absorb moisture from the air.

Finally, unless you are drying enough to fill a kiln, your wood will be held until enough has accumulated. Or you might have to wait until enough lumber of the same or compatible species has accumulated. In addition, pre-drying could add a month or more to the time involved. Getting your lumber kiln dried is not like having your dry cleaning done: there is more than a one- or two-day turnaround.

There is one other option: you can purchase a kiln or build your own. Names of manufacturers of commercial kilns are available at www.drykilndirectory.com.

Some companies sell publications that include kiln construction plans and operating instructions. Plans for small kilns are available in several woodworking magazines.[41]

Air drying

Kiln drying could take weeks, even one or two months. Even so, it is still faster than air drying which routinely takes months, depending on the species, thickness of the lumber, how it was cut (plain-sawn dries faster than quarter sawn), the local climate, and the time of year. There are other advantages to kiln drying beside the time it saves. As I point out above, the operating temperatures of kilns are high enough to kill insects and fungus. Kilns also crystallize or harden the resin in pine. There is also the argument that kiln drying yields lumber with fewer drying defects such as splitting and cupping.

So why would anyone air dry their lumber, given the apparent advantages of kiln drying?

The short answer is that while air drying is slower it is also less expensive and, when properly done, can yield the same quality wood as kiln drying. With a few exceptions, I have used only air-dried lumber cut from urban trees for all of the projects I have made since the mid-1990s. The only lumber I have lost due to air-drying defects consisted of several wide ash boards that contained the tree's pith, and a stack of oak that I waited too long to sticker. The former probably would have split in a dry kiln as well and the oak would have turned out fine had I stickered it in time.

Air drying takes more time than kiln drying, especially in the winter when the temperatures are lower (in fact, drying stops below freezing). Still, air drying does not take as much time as you might think. From my own experience, I have concluded that most lumber can be dried in six months or less, depending on whether outdoor drying begins in warm or cold months. This longer, variable schedule would be an expensive bottleneck for commercial lumber manufacturers but, I believe, is not a comparable problem for modest efforts to harvest urban timber.

To understand how air drying works, we must begin with the meaning of the term itself. The "air" in air drying always refers to outside air.[42] Throughout North America, outside air drying will bring the EMC of wood somewhere below its fiber saturation point (FSP) of 25 to 30 percent. The more exact EMC varies widely.[43] As you can see in Appendix H, which gives the EMC of outdoor wood for selected cities throughout the United States, in June wood in Phoenix will dry down to an estimated 4.6 percent (the result of Phoenix's well-known high temperatures and low humidity). This is below the 6 to 8 percent typically produced by dry kilns. In the same month, wood in Los Angeles, a day's drive from Phoenix, would dry down to about 15 percent. And in the same month in Indianapolis, the nearest city to my own, wood would dry to an EMC of almost 13 percent. In January, the EMC for Phoenix is just over 9 percent, about 12 percent in Los Angeles, and 15 percent in Indianapolis.

If I were building something for outdoor use—public furniture such as a picnic table, a bus stop, or a park bench—I would use lumber air dried to the approximate EMC for that urban area. As you can see in Appendix H, EMC varies across the country and changes everywhere from season to season. This means that in constructing the table or bench, I must make allowances for seasonal expansion and contraction. If I fasten every piece together rigidly with no leeway for movement, then the piece might pull itself apart like the sofa table I described above, which popped its oak top one winter day. Unless I am in a hurry, waiting for the wood to air dry down to its EMC is a much less expensive alternative to kiln drying. Unless I am especially concerned about insects or fungus, there is little reason to justify kiln drying to 6 or 8 percent and then storing the wood outdoors and waiting until, for most urban areas, it absorbs water up to its EMC.

If I am building something for indoor use, then my concern is for the relative humidity and related EMC inside the home or building. Assuming the structure is heated and air-conditioned, indoor EMC will be lower than it is outdoors. Indoor relative humidity will vary, as will EMC, but not as much as much as they would outdoors. That indoor EMC is lower and varies less is the reason I believe a distinction should be made between outdoor air drying and indoor air drying.

I dry my wood in two basic stages: first, outside down to the outdoor EMC for my area. Using an electric moisture meter—an essential piece of equipment that you must have—I wait until the MC of the lumber has reached a stable percentage for the season. I check boards in the middle as well as on top, since the middle boards might not dry as quickly. Once EMC is stable for the stack, I move the wood into my basement workshop for the second and final stage: indoor air drying. The wood is ready to use once it reaches its second lower EMC.[44]

Upon hearing my description of air drying, one specialist on kiln drying told me that I was using my house as a great big unregulated dry kiln. This is true. I do not regulate the humidity and temperature of our home to dry wood, nor to accommodate the MC of kiln-dried wood used in the flooring, cabinets, furniture or the wooden frame of the house. I regulate humidity and temperature to suit our personal comfort: we want to be warm and humidified in the cold months of winter, and cool and dehumidified in the hot months of summer. Everything in and of the house must accommodate our heating and cooling schedule, not the other way around. From my point of view, what counts in drying is not that MC is between 6 and 8 percent, or any other percentage. What matters is that the wood I use has accommodated itself to the conditions that prevail where it will live with us in its final form as framing or furniture. Wood in our homes and offices has to adjust to our indoor living conditions. To be more precise, because wood is hygroscopic, the items we make from it must allow for expansion and contraction in response to variations in the indoor environment we have regulated to suit ourselves.

I have learned from experience that the MC of wood that has been graded and stamped kd (kiln dried) may be well above what it was when it left the kiln. For one project with a close completion date, I pur-

FIGURE 79
Lumber stacked by layer for drying. Each layer is separated by stickers so that air can circulate among the layers. Both ends of every board are coated with a sealant to keep the pace of drying below the stress level of the wood. Notice that quarter-sawn boards are on top and flat sawn are on the bottom.
Computer-generated image by Carey Sherrill

chased red oak from my local lumber center. What I needed came from the center's warehouse, which was neither heated nor cooled. In addition, the warehouse's large overhead doors to the outside were left open during business hours to accommodate pickups and deliveries. The kiln-dried oak that came directly from the warehouse registered about 13 percent MC on my moisture meter. Any moisture that was not absorbed while the wood was stored in the warehouse could have been absorbed during shipping. In any event, I still had to air dry the wood in my shop until it reached the EMC for my house. In this case, and in many if not most other circumstances, the final drying step will be indoor air drying.

In general, the key ingredients to successful outdoor air drying are proper air circulation, protection of the wood while it dries, and sealing the ends of every board.

As illustrated in Figure 79, boards must be arranged in layers with air spaces between each layer and each board, in order for air to circulate over the faces of every board in the stack. Placing stickers at 16-inch intervals from one end of the layer to the other creates air circulation space between the layers. Once a stack is completed, the stickers should line up vertically. Both ends of every board should be supported as well. For stickers, I use pine one-by-one-inch firring strips from the local home center. To avoid staining the boards (called sticker shadow), do not use green wood, or strips cut from tannic woods like oak. Even dry oak strips will absorb water and then, as they dry again, will stain the wood that is being dried.

I retrieve discarded pallets from a nearby local pool and patio supply business, and from a motorcycle dealer. I place the pallets on a concrete pad beneath my deck (which is about 15 feet off of the ground), put down a row of stickers, and then stack and sticker the boards. The following procedures for stacking and stickering should expedite drying and minimize drying defects:

1. Freshly cut boards should be stickered immediately, especially during warm weather. By immediately, I mean within a day or so after cutting. Moisture and warmth between the boards creates an ideal environment for fungal growth. Both also promote excessive end splitting. While at first the fungus only stains the board faces (and can be removed by planing), eventually it moves deeper, ruining the appearance of the entire board. As I mentioned above, the only wood I have lost to air drying was oak I failed to sticker before fungal growth began. So, your cutting arrangements should include a plan for expeditious stickering and stacking.
2. Make sure the pallets are level with one another, otherwise the lumber will dry in a bowed or twisted position. If pallets are not available, use treated boards placed on edge to elevate the stack. If the area is dry and well-drained, as mine is, the stack can be elevated about six inches off the ground or off the pad. If the area is damp, elevation to 12 or 18 inches is better. For damp areas, place a sheet of plastic down beneath the pallets or boards to serve as a moisture barrier. The ground beneath the stack should be stable and well drained, so that the weight of the

wood does not cause a corner or edge of the stack to sink.

3. If you can, shelter the stacks in an old barn, unused space in a garage, or in a well-ventilated utility building. While there are very few old barns in urban neighborhoods, they can be found at the fringes of most cities. You might be able to rent space in one. Warehouse space would work as long as it is well ventilated. With side windows left open, I used my detached two-car garage for several years. Do not store in a space that has little or no outside air circulation such as a closed garage or a storage locker.
4. If the stacks are not sheltered, then, using a carpenter's level, tilt the pallets or boards slightly in one direction for drainage. Even though tilted, boards and pallets should still be level and in-line with one another. Then cover the stack with sheets of corrugated metal or plastic tied down either to the lumber or weighted down with stones. A tarp can be used as long as the stack is not wrapped with it in a way that traps moisture. Being sheltered or covered protects the lumber from rain and snow and, more importantly, from the sun's damaging ultraviolet radiation.
5. If the lumber is all the same thickness, I place the faster-drying flat-sawn lumber at or near the bottom of the stack and the slower-drying quarter sawn boards above them.
6. In each layer, leave a space of about half an inch between the edge of each board and the edges of its neighbors.
7. I place thick boards (more than one inch in thickness, as a rule) together in a separate stack. I double-sticker each layer and leave wider spaces between the boards in each layer. Greater spacing all the way around allows more air circulation among boards that dry more slowly than the thinner ones.
8. Near the bottom and the middle of the stack, I place boards that are marginally thinner than the surrounding ones. As shown in Figure 79, I can pull a board out and check its moisture content. This gives me a measure of the drying rate of the otherwise inaccessible parts of the stack.
9. If children can get to the drying area, then, for their safety keep the stack heights low enough so that they will not topple over and onto anyone climbing on them. Stacks about 5 feet in width by 5 feet in height should be safe. As an additional safety measure, you can strap all the boards in a stack together (be sure to place small pieces of the sticker material beneath the straps to provide air circulation). To discourage smaller children from crawling under the stacks, tack chicken wire around the bases to block the openings. This is especially important if the stacks are more than 6 inches off the ground or the pad. Children can be told not to play on or under the wood—for whatever that's worth!

I trim all split or cracking ends first before stacking. I coat the ends of every board with Anchorseal, the same waxy liquid sealant I use to coat the ends of logs that are not going to be cut right away. Coating the board ends forces water to exit more slowly through the board faces. Slowing the drying rate reduces drying stresses that might otherwise cause splitting and cracking, especially at the ends of the boards. Coating prevents splitting and cracking but will not heal the breaks already there. Even with end coating, existing breaks might travel the length of the boards.

After the boards have been stickered, I periodically measure MC and keep a record of dates and measures. Over time, this gives me a more accurate idea of the pace, by season, of how quickly my wood dries. In general, I know the wood will dry more quickly during warm months than it will during cold months. Thicker wood dries more slowly than thinner cuts, and quarter-sawn dries more slowly than flat-sawn.

Once the wood has reached its outdoor EMC, I move it indoors. While it might not be necessary, I sticker the wood that I've brought inside as well. From my experience, stickering indoors speeds final drying. It also further minimizes the likelihood that lingering moisture will foster fungal growth (even though fungi do not grow below a consistent MC of 20 percent). I monitor indoor drying just as I did outdoors. When the wood has reached its indoor EMC, it is ready to use. However, if I resaw a thick piece of stock into thinner boards (to bookmatch the pieces for a tabletop or a cabinet panel, for example), I wait several days after resawing before using the wood. Even though the moisture meter readings indicate that the wood has reached its EMC, resawn boards have warped on me. So, I wait a few days for any lingering tension to reveal itself.

I have stored freshly-cut lumber in my shop, skipping the outdoor drying stage. As long as the board ends are coated, the wood dries more quickly than it would outdoors and, in my experience, with little or no additional defects. When both hardwoods and softwoods dry, they not only lose moisture but also give off VOCs, volatile organic compounds. Of those given off, methanol and formaldehyde are the two which, in sufficient airborne quantities, could pose health problems. An advantage of outdoor air drying first, then finishing inside, is that most of these two VOCs will have outgassed outside. What little remains should not pose an indoor hazard. Skipping the outdoor stage and putting wet wood (still above its FSP) in your home or building means that the methanol, formaldehyde and moisture are given off indoors. Even so, modest quantities of wet wood drying inside should not pose a hazard either.[45]

Finally, as noted above, one of the advantages of kiln drying softwoods that contain resin (such as pine, larch, spruce and douglas fir) is that the resin is hardened at temperatures that cannot be reached by air drying (resin crystallizes after 24 hours at 165 degrees F or above).[46] Air-dried resin will eventually dry and harden somewhat, but even then it remains pliable. In addition, while the lumber is drying, the resin oozes from the resin canals onto the faces and edges of the boards, and remains sticky for many months.

I have a stack of pine in my shop that has reached its EMC. The resin is no longer sticky but is pliable. If I used this wood for indoor furniture, I will first remove the resin by hand from the edges and faces of the boards before running them through my saw, planer, and jointer. Resin stuck to the cutting edges and sides of saw blades and planer and jointer knives will burn from the friction of cutting. Once burned, it will tenaciously stick to the blades and knives, reducing their cutting efficiency and ultimately dulling them. Once I have finished constructing and sanding a piece from this wood and am ready for final finishing, I will apply shellac first as a sealer coat to prevent small amounts of residual resin from bleeding through the final finish. More work is required in air drying and then using softwoods that contain resin as compared to using the same woods that have been kiln dried. However, I have purchased kiln-dried softwoods from lumber stores containing resin that was not fully hardened. Overall, in my experience, some extra effort will be needed to work around resin, regardless of how the wood was dried.

Summary

After felling and bucking urban trees into logs comes the task of sawing the logs into lumber and then drying the lumber to a MC appropriate for either indoor or outdoor use.

Logs must be skidded, loaded, and hauled safely. Winches can be used to move logs to the loading or cutting site. Whether dragged across the ground or hauled in a buggy, winches, chains, cables, pulleys and buggies must not be pushed to their load capacity. Special care must be taken to protect everyone involved from injuries caused by sudden breaks in cables and chains. Once loaded, logs and lumber must be securely strapped together and to the truck bed or trailer. While sawing logs is generally a safe task, muscular and joint injuries are possible from too much lifting. All those involved should pace themselves by taking breaks.

Much of the time and expense that go into harvesting urban timber comes from moving the logs and lumber from one place to another. The cost per board foot of lumber goes up each time the wood is moved. At most, it can be loaded, unloaded and hauled three separate times. The logs would be loaded and hauled from the felling site to the mill where they would be unloaded. Once the logs have been sawn, the lumber would be loaded and hauled to a kiln for drying (where it would have to be stickered as well). After drying, the lumber is then loaded once again and hauled to a storage place or a woodshop where it would be unloaded for the third time.

By contrast, moving time and expenses can be eliminated when the logs are sawn, air dried, and stored where the trees were felled. In the best of all possible worlds, the sawmill can be set up right next to the logs so that skidding requires no more than simply rolling the logs up to the mill. When logs and the site are prepared in advance, time and expense are limited to positioning the logs for loading onto the mill, sawing lumber, off-loading boards as they are cut, stickering and stacking, coating the ends of the boards, and waiting for the lumber to air dry.

Volunteers can do the first five tasks either for a share of the lumber for their own use or for a share to be donated to a community or educational project. Students can be hired if there are not enough volunteers. High school and college athletic teams often work to raise money to support their sports. I have hired members of my university's rowing team to load and unload lumber.

When logs do have to be hauled to a mill, the least expensive approach is to pay the tree service or park service an extra fee to do the hauling. Also, you might have to pay extra to have the trees bucked into saw log lengths instead of the usual shorter pieces cut strictly for ease of loading. The lumber will have to be loaded and hauled to the shop, a storage facility, or to the kiln, which adds to the cost per board foot. Whether all these steps are necessary depends on whether you plan to dry and use or sell the wood yourself or sell it wet or air dried to someone else.

If you are selling wood to commercial buyers, then you need to be aware of the standards and language used in the commercial markets. The NHLA (National Hardwood Lumber Association) sets dimensional and appearance standards for hardwoods while the mainly structural standards for softwoods are set by the ALSC (American Lumber Standard Committee). The AWI (Architectural Woodwork Institute) sets appearance standards for both hardwoods and softwoods that differ from those of the other two organizations. If you plan to market your wood to commercial buyers, you need to know which standard the buyers follow and, therefore, what they expect. Prices are available from *The Weekly Hardwood Review*, the *Hardwood Market Report*, and Random Lengths' *Lumber and Panel Market Report*. These three organizations can help find local grading assistance.

When you are cutting for yourself, you must choose both dimensions and types of cut. Boards can be cut flat, rift, or quarter sawn. Each has a different appearance and drying characteristics. For example, the appearance of quarter-sawn oak is prized even though it is somewhat more difficult to saw and takes longer to dry than flat-sawn oak.

You must also decide whether the quality of what you are cutting is sufficient for furniture and high-end crafts or whether the wood is better suited to outdoor uses such as trail steps or landscaping ties. A potentially significant outlet for lower quality wood is the pallet and container industry that uses some softwoods and a substantial amount of the annual hardwood output nationwide.

Drying is the final step. Lumber can be either air dried or kiln dried. Air drying takes more time but is less expensive, especially for small amounts of lumber. Either way, boards dried too rapidly develop defects such as cupping, cracking and splitting. I air dry lumber outside down to its FSP, which is about 25 percent MC. I stack and sticker the boards in layers, end-coat them to limit splitting and other defects, and cover the stacks to prevent damage by ultra-violet radiation from the sun. Once the wood reaches its FSP, I bring it in indoors for the final indoor air drying. Whether the wood is air dried or kiln dried, it must come to its EMC with the relative humidity of the setting where it will be used. A moisture meter is essential for keeping track of drying rates and moisture levels.

4 WHO IS HARVESTING URBAN TIMBER?

Introduction

Previous chapters have established the potential value of urban trees as a source of lumber, described safe methods of converting the trees to logs and then to dry lumber. This chapter presents eight stories, five in detail, about those who are acting on the idea of harvesting urban timber.

The first relates my own personal effort to make use of one specific tree. The second story is about a company that is not part of the tree care or lumber industry but strives to make the best use of its large stock, even one tree at a time. The third describes how, at the state level, the right woman in the right position can make a big difference in the way urban trees are used in education, even when all the trees come from just one place. The fourth story is about how another woman in the right place single-handedly converted the trees in her small city into a local resource. The final story concerns a southern California tree-care company that is trying to create a viable business from its commitment to making the best use of urban trees it removes. The chapter ends with brief descriptions of the efforts of three men, two in California and one in Utah, who were among the first to reclaim discarded trees.

The 500-Year-Old Bur Oak[1]

"Black John" and the Bur Oak

In the middle of the nineteenth century, John Newton Gatch, a riverboat captain whose long black beard earned him the nickname "Black John", returned to his native southwest Ohio to live. In 1850, with money earned on the river, he purchased a 180-acre farm and a stone farmhouse (built about 1820) in what is now Milford, Ohio. He acquired the acreage, the house, and a very large bur oak tree in front of the house (Figure 80). At that time, the oak was estimated to be about 350 years old. According to the Gatch family history, it was the large tree that drew Black John to this particular farm. In the life of a riverboat captain, the landscape changes every day; nothing is constant except the motion of the river and the boats. Perhaps the big oak represented permanent anchorage for a man whose life had taken him from one place to the next along the Ohio and Mississippi Rivers. He remained on the farm for the rest of his life, raising a family and working the land.

Such was his reverence for the oak that he would not allow anyone to eat beneath it for fear that salt would fall from the table and damage the tree's roots. On his death in the summer of 1891, his friend and then Governor of Ohio, John M. Pattison, delivered Black John's eulogy next to the casket as it sat beneath the tree.

The Oak Falls

For almost 150 years from the time Black John purchased the land, the oak was important to each generation of the Gatch

FIGURE 80
John Newton "Black John" Gatch (beardless) seated beneath the bur oak with his wife Georgiana (left) and their daughter Bessie (right)
Circa 1870 by unknown photographer

family. Then, in 1996, a spring storm blew through southwestern Ohio and knocked almost the entire tree to the ground. Lewis Gatch, a great-grandson of Black John, asked me to inspect the limbs and trunk and determine what would be required in order to saw them into lumber.

I have seen the world's largest trees in the redwood groves and forests of northern California. Pat and I once spent most of a Sunday in France's Loire Valley searching for the largest and oldest oak tree in France (about 350 years old). Even these experiences did not fully prepare me for what I saw when we visited the Gatch farm to see this particular tree.

Though most of it was lying on the ground, I was still awed by the size of the stump and the limbs. The trunk had been roughly 10 feet in diameter at breast height. The largest limbs were almost 3 feet in diameter, which is the size of the trunks of mature oaks growing in this area. This was the largest tree I had ever seen outside northern California. It was both larger and older than the oldest French oak.

For the current generations of the Gatch family, this was not just the loss of tree but more like a death in the family. It had been a silent participant in family gatherings and celebrations since 1850. Children had climbed its trunk and swung from its limbs, successive generations of the family had picnicked beneath it on summer days (being careful not to spill salt), and had even been married beneath its green canopy. As each generation succeeded the next, the oak stood as the enduring link between them, the center of family memories.

This was an unusually old tree even for a species of oak known for its longevity.[2] From a broader historical perspective, it had been a sapling about the time Columbus set sail for the New World. The spring storm in 1996 took apart in minutes what had survived and grown for five centuries. During that time, a countless number of storms and even tornadoes passed through this part of southeastern Ohio, more than a few passing directly over the oak itself. It withstood them all. This particular storm was not all that severe. So why had the wind knocked it down?

I realized what had happened when I inspected the trunk. Over the past half century, the heartwood had slowly rotted away. The growing void had been repeatedly filled with concrete. Finally, the thin layer of sapwood and what was left of the heartwood could no longer support the weight of massive limbs that extended, in each direction, more than 50 feet from the trunk. In this last storm, the tree collapsed. In the midst of fallen limbs and debris, the concrete rose out like a headstone from the shattered trunk.

From Limbs to Lumber

Because the trunk was hollow, only the limbs were sawn into lumber. The boards were then dried in a kiln at a large commercial sawmill in central Ohio. When I drove up to take delivery, the mill owner told me that the wood was not very good, more suited, he thought, for pallets or firewood. I told him that the tree was about 500 years old and of great sentimental value to the family. He shrugged and repeated his opinion, pointing out that lumber from tree limbs is considered stress wood and would be very difficult to work. The boards were all neatly strapped together and ready for loading, so I did not get a close look at them. Those that I saw looked straight and flat.

Only after I got home and began unloading the truck did I get a close look at each and every board. As the last one passed through my hands, I fully understood the mill owner's comment and the challenge this wood posed. Most of the boards had cupped, bowed, and checked, contorting themselves in every possible way during drying. I realized then that whatever I made from this wood, I had my work cut out for me.

From Lumber to Furniture and More

At Lewis Gatch's request, the first project was a large dining table for the old stone farmhouse. We selected a trestle design with a 9-foot top that I patterned after a table I had seen Norm Abram build on *The New Yankee Workshop.* I wanted to finish by late November so that the family could have their traditional Thanksgiving meal on the table in the farmhouse.

My first challenge was to find four boards at least 8 feet long, and 8 to 10 inches wide, for the tabletop. I ended up standing the longest and thickest boards on end against my driveway wall and then moving them back and forth, swapping one on the wall for one in the pile, until I finally found four boards wide enough, straight enough, and flat enough to make the top. Out of about 200 boards totaling more than 1,100 board feet, only four boards met these requirements.

The grain was unique and beautiful, consisting of wavelike patterns, soft-pointed polygons with curving lines and

FIGURE 81
Unusual grain pattern(s) created by bookmatched boards.
Photo by Sam Sherrill

rolling circles. Bookmatched pieces created Rorschach patterns that looked like animal faces from one angle and devilish human faces from another (pictured in Figure 81). The price for this unique grain was a problem known as tear-out: small chunks of wood were ripped from the board faces no matter how lightly I passed them through my thickness planer. More divots of wood were torn out as I used my router to create straight, flat edges for gluing. The more I worked this wood, the more it worked me. In frustration, I often groped for new combinations of old expletives. As part of his trip to Cincinnati in 1998 (described in the Introduction), Norm Abram of *The New Yankee Workshop* visited the farmhouse with me. I felt better when he pointed out that this wood would challenge the most skilled woodworker.

The four long boards bowed a bit more after drying in my basement workshop for several weeks. I glued them together one at a time using every clamp I own. Using oak splines in the edges of the boards, I clamped one bending up to one bending down, in the hope that each would offset the other so that, when I finished, the top would be flat. The finished top was flat and, at about 80 pounds, was also heavy. I eliminated tear-out from the planer by running the top through a 52-inch drum sander at a local millwork company. Again, light passes were essential, this time to avoid burning the wood. Fortunately, the two trestles, and the stretcher that holds them together, were easier and took less time to build than the top.

I managed to complete construction by early November. After applying four coats of finish, I delivered the table (Figure 82) to the farmhouse right before Thanksgiving. Afterwards, Lewis Gatch told me that he and his family were grateful to have had the table in time for the family gathering. For them, it was a tangible reminder of an

FIGURE 82
The finished trestle table. A 6-inch breadboard end at each end of the table brought its total length to 9 feet.
Photo by Sam Sherrill

important part of their family history. I was very pleased to have created something of deep personal value for the family, and also to have worked with wood from this very old tree.

Since making the table, my son and I have made a total of 20 pieces for members of the Gatch family. We designed and made a desktop box, a fireplace mantle, several rocking horses for the newest generation, a kitchen table, and more than a dozen living room tables. Cathy Gatch, a great-great granddaughter of Black John and a talented potter, created two sets of nearly identical tiles depicting the old stone farmhouse and the tree. As pictured in Figure 83, we used them as inlaid tops for two end tables made from the oak.

From what I have learned about him, I believe Black John would be very unhappy to know that the tree he so revered has fallen. But I also believe that he would be pleased to know that, in its new forms, it survives and is still in the hands of his family and will be so for generations to come.

The Trees of Biltmore

The Biltmore Estate and Inn

Biltmore Estate in Asheville, North Carolina was constructed between 1889 and 1895 for George Washington Vanderbilt, the grandson of "Commodore" Cornelius Vanderbilt. Modeled after a 16th-century French Renaissance château, the building was designed by architect Richard Morris Hunt, and the grounds by landscape architect Frederick Law Olmsted, who also designed New York City's Central Park. Described as America's largest home, it is large by any standard, containing 250 rooms including 35 bedrooms and 43 bathrooms. Situated on 8,000 acres, the estate, its gardens, and winery are open to the public. To accommodate visitors, the 213-

FIGURE 83
Inlaid tiles depict the bur oak and the stone farmhouse.
Photo by Sam Sherrill

room Inn on Biltmore Estate was built. The Inn opened in the spring of 2001.

The Lone Black Walnut

The single black walnut tree pictured below was growing at the edge of a planned parking lot and driveway for the new Inn. The initial plan was to establish a perimeter around the tree, which would protect it during construction and also allow it to continue growing thereafter. However, given the extensive nature of site preparation, arborists judged that, in the near future, the tree would succumb to construction stress. It was taken down, quarter sawn, and then kiln dried by students in the Natural Resources Program at nearby Haywood Community College. A local millwork company cut and planed the boards that were then used for flooring in the Inn's private dining room (shown in Figure 86).[3]

The fact that saving the walnut tree was the first option considered reflects the overall approach the Estate takes toward its forest resources: to harvest and sell, exchange, or use trees when they must come down, but otherwise preserve and maintain them whenever possible.[4] In short, the Estate's forests are managed from an arboricultural, not logging, point of view: the trees are treated as part of the Estate's gardens, not as a crop to be harvested and sold.

But even with the best care, trees still

FIGURE 84
A black walnut tree growing at the edge of the site for the Inn on Biltmore Estate.
Biltmore Estate, Asheville, NC

must come down. Storm damage, drought, old age, disease, and insect infestation are the typical reasons at Biltmore. To make the best use of the wood, the Estate has an agreement with a local Wood-Mizer® mill owner to saw the logs and evenly divide the lumber. The Estate uses its share for everything from flooring, paneling and construction of storage buildings, to temporary bridges. At present, it has accumulated an inventory of lumber to meet its own needs.

There are a substantial number of white pines ranging in age from 90 to 110 years old growing on the Estate grounds. At present, many of them are infested with the Southern pine bark beetle, an infestation that is fatal to the trees. The diseased and dead pines in the areas around the gardens are felled and sawed into lumber, eliminating what would otherwise be a disposal expense. Because the standing dead and dying pines contain an unknown amount of rot and blue stain, they are a

FIGURE 85
Lumber cut from Estate oak trees was used for paneling and trim in the Inn library.
Photo by Patricia Sherrill

FIGURE 86
Random-width walnut flooring in the private dining room of the Inn on Biltmore Estate.
Photo by Patricia Sherrill

risky purchase and not very valuable to log contractors. However, once felled and sawed into lumber, the market value would be between $500 and $600 per thousand board feet. Sawing adds value by converting the logs into lumber and by eliminating the risk of buying trees with rot and blue stain. At present, the Estate uses all of the pine lumber it generates.

From the Grounds of the Governor's Residence

The Indiana Governor's Residence

In 1973, the State of Indiana bought a privately-owned home not far from downtown Indianapolis to serve as the Governor's Residence, the sixth since Indiana became a state in 1816. With poured concrete floors 1 foot thick resting on concrete walls, this very substantial English Tudor home sits on 6.5 acres populated by a variety of tall and elderly hardwoods. An inventory and assessment revealed that many of these trees have grown frail with age and are no longer capable of withstanding the storms that sweep through the middle of Indiana every spring and summer. That they might topple over anywhere on the grounds poses a problem: that they might land on the Residence makes them a significant hazard and therefore candidates for removal.

Lumber for Indiana Schools

Frank O'Bannon was first elected governor of Indiana in 1996 and then again in 2000. In early 1997, his wife, Judy O'Bannon, was meeting with the Residence director when its usually quiet atmosphere was disrupted by the sound of chain saws.[5] A tree service had been hired to fell several large cherry and walnut trees that had been designated as hazardous. Mrs. O'Bannon recalls asking a young man with a chain saw what would become of the trees. His answer was that he guessed they would end up in a nearby landfill. Her response was that they would not!

From her own experience rebuilding a log barn in southern Indiana, where the family home is still located, Mrs. O'Bannon is familiar with the difficulty of finding just the right wood. Seeing good wood go to a landfill just was not an option. She said that at that moment she did not know what she was going to do with the logs but she was determined that they would not be thrown away. In the back of her mind was the possibility that the wood could be used by students to make furniture and crafts, although she did not yet know exactly how to make that happen.

An initial step was finding a way to have the logs sawn into lumber. Through a state forester, she discovered that the Wood-Mizer® Company was headquartered right there in Indianapolis. A company representative was contacted and agreed to saw the logs into lumber. The wood was stored and air dried.

There are three primary themes that direct Mrs. O'Bannon's own efforts on behalf of Indiana. Education was a top priority for the O'Bannon administration from the outset. In support, Mrs. O'Bannon visited high schools throughout the state where she learned from wood-shop teachers that the shift toward technical training, and the accompanying loss of federal and state funds, increasingly deprived students of the opportunity to learn how to work with wood. Both the skill and the opportunity for artistic expression were being lost. The teachers asked for her help. Secondly, she views the Governor's Residence as a unique public place—the State's living room—where the talents and accomplishments of

FIGURE 87
Governor O'Bannon (middle) and Mrs. O'Bannon admire clock made from one of the Residence cherry trees by a high school student.

FIGURE 88
Mrs. O'Bannon runs her fingers over the silky smooth surface of a box made by a student from the Indiana School for the Blind. She takes it with her in trips across the state to use as a tangible example of how wood that could have ended up in a landfill instead became a finely crafted box in the hands of students who could feel its promise.

its citizens could be recognized, honored and displayed. She felt that tokens of recognition awarded by the state should be more than the usual engraved plaques or printed certificates. Thirdly, community building is a central and re-occurring objective in most everything she does. Making good use of community natural resources is an essential part of that effort.

Festivals of Growth

Her desire to help wood-shop teachers, to make the Residence a welcoming place, to adequately recognize achievements, and to demonstrate the best use of a scarce community resource—all of these gave rise to the idea of a festival centered on urban forestry and conservation. The lumber sawn from Residence trees was distributed to about half a dozen high schools, including the Indiana School for the Blind. Then,

FIGURE 89
Mrs. O'Bannon tries on a harness and ascending rig used by arborists to work in trees.

FIGURE 90
Mrs. O'Bannon saws a board on a Wood-Mizer® band mill.

in the spring of 1998, students, teachers and principals gathered at the Community Forestry Fest on the grounds of the Residence. This was the first celebration of the annual event now known as the Festival of Growth, which includes the Indiana Arborists Tree Climbing championships, recreational tree-climbing for children and adults alike, log-sawing demonstrations, and displays on urban forest conservation.

Students attended the first festival to show what they had made, and to receive recognition for it. While most items were small—boxes, clocks, trays, or wall shelves made from cherry and walnut—several large pieces were made as well, including a table, a blanket chest, and a sideboard. These are either in use or on display in the Residence. Numerous pieces have been given as gifts to a variety of people, ranging from state employees in recognition of outstanding service, to special visitors from other countries. These are unique examples of how Indiana makes use of its resources. Participants and their schools received plaques handmade from the same lumber.

Lompoc Lumber

Lompoc, California

Located in Santa Barbara County, Lompoc is a small city with a population of just over forty thousand. Just off of the Pacific Coast Highway (Highway 1) nine miles inland from the Pacific Ocean, it is about one hundred and fifty miles north of Los Angeles and about fifty miles north of the city of Santa Barbara, just southeast of Vandenburg Air Force Base. There is a federal prison at Lompoc and the hills to its south are mined for diatomite, the raw material used to make diatomaceous earth.[6] Its name, which is also the name of the surrounding valley, comes from the Chumash Indian word for small lake or lagoon. The Chumash have occupied this general area for the last three to five thousand years.[7]

The Band Mill Experiment

As mentioned in Chapter One, the CDF (California Department of Forestry and Fire Protection) owns five Wood-Mizer® band mills and four transportable kilns that can be borrowed by local governments, non-profit organizations, or private firms interested in sawing lumber from urban trees that would otherwise end up in municipal landfills. Making this equipment available for up to a year without charge was one of CDF's contributions to the statewide effort to reduce by 50 percent the amount of solid waste entering municipal landfills by the year 2000 (using 1989 as the base year). The legislation mandating the effort (AB 939) placed the highest priority on reductions at the sources of waste.

In an effort to reduce green waste in Lompoc, Cindy McCall (Figure 91), manager of the Parks and Urban Forestry Division, had her crews cutting and carving benches from logs using chain saws (Figures 92 and 93) while either cutting the rest for firewood or grinding them into mulch. She also donated some to local woodworkers.[8] However, she realized at the time that these efforts would not be enough to stem the flow of about 3,000 street and Park trees designated for removal by the year 2000. Assuming the Division could handle the volume, her only choice would be to cut and grind many good quality logs. Having heard of the CDF offer to make mills and kilns available, in 1998 she applied for and was the first to receive one of their band mills.

Felling, limbing, bucking, grinding, hauling, and tippage fees at landfills are the expenses of urban tree removal. Even when public properties are used as disposal sites, there is the opportunity cost of using the land for disposal instead of other, possibly more valuable, purposes. Some of the expenses can be recovered when the trees are either ground up for mulch and used in the parks and public gardens or sold as firewood. Parks and urban forestry departments are also major buyers of wood for outdoor furniture, shelter houses, steps, railings, parking lot blocks, and other parks-related infrastructure. They are in the ironic position of either paying to dispose of or giving away urban wood on the one hand, and buying substantial amounts of lumber on the other. Cindy saw the CDF band mill project as a means to several ends. She expected it to reduce the tippage fees paid for the disposal of green waste in Lompoc's only landfill, fees that could rise dramatically when the 3,000 trees scheduled for removal started coming down. She knew the mill should help the city meet the state's goal of a 50 percent waste reduction (at the source). Lumber produced by the mill would reduce the Division's purchases for repair and replacement of public furniture, and allow it to use more durable wood than they could afford to buy. Finally, in the hands of local woodcrafters and Division

FIGURE 91
Cindy McCall, manager of Lompoc's Parks and Urban Forestry Division.

personnel, it would create opportunities for the imaginative use of reclaimed wood in Lompoc's public parks and private spaces.

The Results: Reduced Expenses, Better Wood, and More Imagination

Pat and I visited Lompoc in mid-1999 to meet Cindy and to see firsthand how the band mill project was working.[9] In general, it helped her Division achieve the ends she hoped it would. A two-man crew quickly learned to use the Wood-Mizer® mill and before the first year was up had sawn lumber from about 130 trees that would have otherwise ended up in the local landfill.

This saved the city about $40,000 in annual tippage fees, an amount much greater than the mill's operating costs of blade-sharpening, replacement, fuel, and engine maintenance (in fact, it about equals the current price of one of the company's largest, fully equipped mills). The manager of Lompoc's Solid Waste Division estimated that the project reduced by 30 percent the total volume of all solid waste going to the landfill, thus substantially helping Lompoc meet the state's 50 percent reduction mandate. Of

FIGURE 92.
Park bench carved by chain saw from a whole log.
Photo by Cindy McCall

FIGURE 93.
Park employees finishing a skateboard bench for the city's park.
Photo by Cindy McCall

course, not all the trees taken down could be sawn into lumber. The unusable trunks, as well as tree limbs and leaves, were mulched. However, Cindy worked out an arrangement with the manager of the city's water purification plant and its Waste Management Board to combine the mulched green waste with the calcium-based sludge from the plant to create a landfill cover. Using the mulch and sludge blend means that the slopes of the landfill itself do not have to be cut for soil to use as cover, thus saving valuable landfill space.

The Parks and Urban Forestry division now produces all of its own lumber, freeing up the funds that had been budgeted for these purchases. These funds can now be reallocated to other purchases such as playground equipment, and sod for play fields. In fact, not only were these purchases eliminated, but also the lumber that the division produced for itself was better than any they were able to afford

FIGURE 94
CDF mill used to saw lumber from Lompoc's trees.
Photograph by Cindy McCall

FIGURE 95.
Memorial bench adjacent to the Lompoc Public Library. Dedications can be laser-engraved on the benches.
Photograph by Cindy McCall

on that budget. Shamel ashes were among the trees the crew first cut. The highly durable ash lumber was used to repair and replace picnic tables and park benches that had been made from the much less-durable pine that the division had been buying. In addition, 3-inch-thick ash boards that cost about $120 to saw were used as replacement flooring in a local community center: 1-inch boards purchased from a lumberyard would have cost between $500 and $900.[10]

Ash and other woods are also being used for bleacher seating, indoor paneling, doors, shelter houses, and park signage. By using their own more-durable wood, the city not only eliminated its immediate purchase expenses but also reduced its longer-term maintenance expenses as well.

For a $1,000 fee, the division constructs memorial benches from ash and places them in parks and other public places. The benches cost about $200 dollars in time and materials to build. The one pictured in Figure 95 is adjacent to the local library.

The basic idea of using instead of discarding urban trees, as perhaps tangibly represented by the mill, was the source of inspiration for some creative work as well. Among the more imaginative efforts are the playhouse pictured in Figure 96, and the finished tree trunk with attached limbs shown in Figure 97—used as a design element in the entry hall of a private residence.

The division has begun accepting trees from private property thus lowering, from yet another source, the flow of green waste into the landfill and thereby reducing the often-substantial tippage fees that would otherwise be paid by property owners. Most recently, the division acquired a substantial number of black walnut trees that were dying. Given the value and intrinsic beauty of walnut, lumber from these logs is being reserved for use as paneling and architectural elements in new city buildings.

Cindy and the division are not focused solely on harvesting public and private timber. She is currently the Executive Director of Santa Barbara County ReLeaf, a member of the California ReLeaf Network. Through local communities and volunteers, ReLeaf works to preserve urban forests throughout the state, by planting new trees while protecting and maintaining existing ones. Looking ahead, the Division is planting black acacia, a fast-growing tree well suited to the Lompoc climate which, when mature, will provide durable lumber for future public projects. This is consistent with the view of CDF's Eric Oldar, mentioned in Chapter One. Eric advocates planting economically valuable trees and then harvesting them while they are still usable.

Owning a Mill

Harvesting public and private trees raised the question of whether Lompoc should purchase a mill and do its own sawing, or whether that task is better left to individual mill owners or tree-care companies. For Lompoc, or for any local parks or waste management department, this can be a difficult question to answer.

From the narrow but immediate perspective of the budget, the most pressing question is whether the department can afford the initial outlay for a piece of equipment that costs between $20,000 and $40,000. The budgets of these types of departments in local government are often closely tied to their narrowly defined missions. Buying something new like a sawmill, which is somewhat outside the department's traditional mission, means either new money must be found or else existing services must be cut. For the department to make such a purchase without disturbing its budget or its sense of mission, it must either find a substantial

FIGURE 96
Is it a play tree house or a tree playhouse?

FIGURE 97
A hall tree.

budget savings, get a one-time allocation from some other level of government, or be awarded an outside grant. However, even then, the prevailing sentiment may be that the mill should be productive enough to pay its own way, including all labor costs.

Larger issues also enter the decision. From a political perspective, some believe that any goods or services that can be provided by private companies should be left up to them, and that government should contract out whenever possible. Mill owners concur by arguing that a government department sawing public and private trees, and selling any part of the product. is unfairly competing with their own businesses.

In Lompoc's case, the budget savings would cover the purchase price of a new Wood-Mizer® mill. However, Cindy compared the costs, especially the labor costs, of owning and operating the mill in-house, with the price of contracting-out. Given the difference in hourly wages and physical productivity, she determined that the city would be better served by writing a contract with a tree service company that saws logs into lumber. The contract she negotiated calls for the company to saw the logs supplied by the city, and share the lumber between them. The company markets its portion, while Lompoc uses its lumber for public projects.

The way in which trees are treated, once they come down, has permanently changed in Lompoc because the way they are viewed has permanently changed. What used to be a costly trash disposal problem has become a renewable community resource. Everyone is winning, including the city's trees.[11]

West Coast Arborists, Inc.

From Tree Care to Lumber

West Coast Arborists, Inc. (WCA) started in 1972 as a small tree care business in Anaheim, California. Incorporated in 1978, it has since grown to a full-service tree care company with 350 employees, 80 trucks, and tree care contracts with 122 cities and municipalities throughout southern and central California. Lompoc is one of those cities.

According to Jeffrey Melin, the recycling coordinator for WCA, the company removes thousands of street and parkway trees every year.[12] Even though the timber they remove is most often diseased, dying, or dead, many of the trees contained much fine wood that deserved a better fate than becoming green waste, firewood or mulch. Cherry, oak, ash, black walnut, pecan, elm, sycamore, black acacia, eucalyptus, pine and olive are among the 30 or so species of trees WCA removes annually.

East-West Urban Forest Products

Like Lompoc, WCA also tried sawing logs into lumber in 1999 by borrowing a Wood-Mizer® band mill from the CDF. Along with the mill, they also borrowed a kiln, and used both for a year to gauge whether sawing and drying lumber from urban trees was a worthwhile addition to their tree service business. Having the opportunity to try the mill and kiln, without having to risk the purchase, plus the support and encouragement they received from Eric Oldar of the CDF, made the difference. WCA bought their own large Wood-

Mizer® mill and a Nyle dry kiln. Also in 1999, WCA formed a partnership with East-West Urban Forest Products, a newly-established company based in San Marcos. East-West buys and then markets the lumber WCA produces.[13]

In addition to selling WCA lumber, East-West also builds park benches and picnic tables (Figure 98) from WCA lumber, many of which are sold to WCA's municipal customers. As pictured in Figures 99 and 100, in the hands of individual woodworkers, the wood has been used to produce unique and finely crafted furniture.

Jeffrey told me that the driving reason behind the lateral move into lumber production was a commitment by WCA and East-West to see the best of the trees put to their best use. Mike Easterling, President of East-West, stated that the company's mission "... is to put the lumber from our urban forests to higher use."[14] The two companies feel that salvaging urban trees would, in a small but significant way, help reduce the logging pressure on old-growth forests throughout the state, and on rainforests throughout the world. As Mike put it, "Every time someone buys urban wood, that's one less tree that has to come down in the rain forest."[15] East-West has acted on this commitment by joining the SmartWood^{cm} Rediscovered Wood Program and obtaining the program's seal of approval (given to companies that reuse and salvage wood).[16] As of 2002, the company is one among about two dozen north American companies to receive SmartWood^{cm} certification. The reality—that this commendable effort would also have to survive as a business—came later.

Scale and Profit

At the start, every business chooses the basic scale of its operation by the amount

FIGURE 98
Park bench made from ash and black acacia by East-West Urban Forest Products.

of equipment it purchases, the size of its plant and offices, and the number of employees it hires. With a single mill and kiln and a growing supply of logs, WCA's output (at prevailing hardwood prices in southern California) exceeds the intermittent quantity of lumber demanded by small retailers and individual end-users. However, they cannot yet physically meet the continuous demand of the commercial lumber market, which requires kiln-dried and graded lumber in tractor-trailer loads.

To find a profitable market between these two extremes, East-West has hired two full-time sales representatives to find new customers. The company sells lumber from its website, and has opened its own retail store in San Marcos. They also sell through retail outlets in San Diego, Anaheim, and Solana Beach, to school vocational programs, prison work programs, and to the parks departments of the cities and municipalities where WCA has contracts. It was Jeffrey's belief that East-West would begin to show a profit by the end of 2002.

Ash and elm are two of the East-West's best sellers with its current customers. Because of the ray fleck grain, quarter sawn oak and sycamore are very much in demand, as are the more exotic woods like highly-figured melaleuca (a member of the

FIGURE 99 CD cabinet made from black walnut and quarter-sawn sycamore.

FIGURE 100 Two-drawer clothes hamper also made from black walnut and quarter-sawn sycamore.

eucalyptus family), quarter-sawn silk oak, olive, black acacia and black walnut.

Is There Profit in Commitment?

So far as I know, these two companies are the first to move aggressively into lumber production and sales, using as raw materials the timber that other urban tree service companies generally treat as waste and as a cost of doing their kind of business. What happens next with WCA and East-West is important, because their experience will help reveal how harvesting urban timber can move from an environmental commitment to a viable business in one of the nation's largest metropolitan areas. Earning a profit in the near future means that East-West will have succeeded in finding a way to survive

FIGURE 101
A load of urban logs.

FIGURE 102
Lumber drying in the WCA kiln.

between the small and large-scale extremes of the hardwood lumber market.

However, as WCA's tree service business grows, so does the supply of good-quality logs that can be sawn into lumber. Increasing the lumber business, making an adequate return on their investment, and keeping their commitment to making the best use of all the best logs means increasing output and sales. There is no shortage of logs, and the sawmill can steadily cut at least 700 board feet per day. At present, the air and kiln drying is the major bottleneck that restrains WCA's production capacity.

To increase output, especially to meet the demand of the commercial lumber market in addition to the demands of their current customers, WCA will have to purchase additional kilns. Tree services in other urban areas could get around this bottleneck (and avoid the expense of buying a kiln) by paying nearby kiln owners to do the drying. WCA cannot do this because, as far as they know, they own the only urban kiln in southern California. They will also have to acquire the services of a certified hardwood grader.

As I pointed out in Chapter One, felling

IN MY PRIME, BEFORE MY TIME

Fifty years I stood and grew,
under skies of vaulted blue.
Then the ax cut me through—
in my prime, before my time!

The birds my bower quickly fled,
and winged the skies or' my head.
City ground's my funeral bed—
in my prime, before my time!

My fate the dump, a wasting space,
per chance ash in a fire place.
Yet I dreamed to be a case—
in my prime, before my time!

Now you hold a work of art,
nor dump nor flames have reached my heart.
Now I live the fuller part—
in my prime, just in time!

© 1992 URBAN FOREST WOODWORKS, LOGAN, UT

FIGURE 103
The Urban Forest WoodWorks explanation, which accompanies each piece, describes the wood it is made from and founder George Hessenthaler's urban tree credo.

FIGURE 104
Box with apricot top, ash sides, English walnut bottom, lined in leather.
Photo by Patricia Sherrill

trees in urban areas is not the same as commercial logging. Given the market demand for lumber and pulp, most trees are identified and harvested based on species, size and quality. By contrast, in urban areas, we must wait for trees to come down for reasons unrelated to demand (for example, storm damage, infestation, disease, hazardous condition, or real estate development) and then make use of what has become available. This is an economically more challenging form of logging: we do not get to pick what we want when we want it—we must be prepared to market what has been already picked for other reasons and by other forces. Orders cannot be filled by simply finding and felling the requisite species and number of urban trees.

Accumulating commercial quantities (10,000 board feet and multiples of this amount) requires waiting until there are enough logs or lumber to sell. The length of the wait depends on the quantity and pace that trees of specific species and quality are being taken down. In most urban areas, putting together a tractor-trailer load of black walnut will take longer than assembling a load of oak, ash or maple. In turn, waiting means having a place in which to store logs and lumber. This is where the challenge lies for WCA and East-West, and for other companies operating in urban areas where property prices are high, and where transportation costs offset the lower prices of more distant properties that could be used for storage, drying and sawing.

Three Pioneering Efforts

Although there are other stories, the following three warrant attention because the three men involved separately, but at about the same time, acted on the idea of reclaiming discarded urban trees.[17] They were among the very first in recent times to do so. Each worked hard to create viable businesses using reclaimed trees.

With a single Wood-Mizer® mill, Dave Parmeter started California Hardwood Producers, Inc. in Auburn, California in 1992. In 1996, he and his partner enlarged their business with $200,000 in revolving credit from the California Hardwood Initiative, a program to develop the state's hardwood industry, funded in part by the USDA Forest Service.

Dave was running a successful business on the site of an old softwood lumber mill in Auburn, producing about 1 million board feet annually, when on July 4, 1997, the entire plant burned to the ground. Though financially devastated by the fire, he did not declare bankruptcy, but instead paid off all of his creditors and, with the help of the Auburn community, successfully revived his business.[18] He credits firewood sales with keeping his company afloat until he could move once again into lumber production. At present, in addition to firewood, his company produces flooring and hardwood lumber. His sources include street trees from cities such as Sacramento, and farmers replanting orchards or selling to developers. None of the trees he uses come from commercially-harvested forests. The lumber is sawn from a dozen and a half different species, including acacia, ash, eucalyptus, locust, madrone, oak, pine, sycamore, walnut, peach, apricot and pear.

George Hessenthaler, journalist-turned-woodworker, is also a pioneer in this effort. He entered the woodworking business in 1977, manufacturing boxes decorated with marquetry inlays. In 1988, he purchased a Wood-Mizer® band saw and began making boxes from trees taken down in and around Salt Lake City, Utah where he was located at that time (he has since moved to Logan, Utah). Also in 1988, he started Urban Forest Woodworks, a company that currently employs up to a dozen people to produce highly-figured lumber, and jewelry boxes and other craft items, from discarded trees removed in Salt Lake City.[19] George began using urban trees for the same reason as others do: he feels that treating them as waste is shameful. In his own words, "It's a crime against nature, and a sacrilege for woodworkers, to dump any hardwood tree, or even to turn a blind eye to the act."[20] George also makes the point that utilizing urban trees should not lead us to fell otherwise healthy trees merely to meet an immediate demand for a specific kind of wood. Again, in his words, we should not become "urban forest tree predators."[21]

Finally, the late David Faison, a founder in 1990 of Into the Woods in Petaluma, California, was among the very first to act on the idea of reclaiming trees and recycling lumber, and among the first to receive SmartWood℠ Rediscovered Wood certification. A former woodworker who grew uncomfortable with the harvesting practices that provided him with the raw material of his trade, his conversion to discarded urban trees was quick and satisfying.[22] His company reclaimed most northern California species, among them madrone, tanoak and bay laurel. From nearby urban areas, he salvaged black acacia, eucalyptus and elm. And, from orchards that were being cleared of aged trees, he acquired apple, pear, plum and olive wood. He also recycled redwood lumber and beams, but considered felling existing redwoods to be unacceptable. Much of what he salvaged was used to make furniture and flooring. The business that David built reflected his deep commitment to conservation and to the responsible use of forest resources. He deserves to be remembered both as a pioneer and as an eloquent spokesman for making the best use of discarded trees and lumber.

5 HARVESTING TREES IN YOUR COMMUNITY

Introduction

We are losing 3 to 4 billion board feet of wood annually to urban landfills throughout the country. Whether, as an estimate, this is a billion too high or a billion too low, the waste is still several billions too many. On the face of it, this is a waste of much good wood and landfill space, as well as money spent on hauling and tippage fees, and on the purchase of wood to meet needs that could have been met by timber that was discarded.

There is no technical barrier that prevents us from using this enormous supply of wood. The machinery and the methods for felling, sawing, and drying lumber from urban trees are available. The demand for wood and wood products continues to grow. We want more of everything, from construction lumber, pallets and skids, flooring and cabinetry, to the finest custom-built furniture and crafts.

Even though the supply and demand are there, the connection between the two has not been made (outside California) in a way that, so far, has led to the widespread startup of tree-reclamation businesses in urban areas throughout the country. Perhaps those who would start these new businesses, such as urban tree-service companies, have considered the idea and have decided that there is too little profit to justify the effort. My view is that this idea has not yet been fully examined by those who might make a business of it. While there is a widespread awareness of wood going to waste, the next step—the possibility of making a business of reclaiming the wood—has not been completely explored.

The simplest conventional wisdom in economics is that new products and services are created solely by the entrepreneurial efforts of individuals and companies willing to incur risks for eventual, and often substantial, profits. While there is some truth to this generalization, it is also true that both government and individuals driven by motives other than profit are also sources of new ideas that are important for our cities and communities. Some can eventually become profitable products and services. Reclaiming urban trees seems to fall into this category: it is an idea that will have to be first promoted by non-profit community work and then supported by local and state governments. Governmental support, related waste-management legislation, and individual experiences together could create a market environment that encourages and helps sustain the business of reclaiming urban trees.

The greatest reduction in the billions of board feet wasted every year will come from the sum of the separately motivated efforts of businesses, government, and community volunteers. While this is the best solution, reclaiming urban trees can and perhaps should start on its own as a community effort. In my view, this effort can be the impetus for subsequent governmental action and the creation of private businesses. What follows is how the community effort can be started.

Getting Started

For amateur woodworkers, deep affection for the material and the tools, for being in the moment when working with both, and for the artistry of the results, is what counts the most. Using wood from urban trees, and the extra effort this entails, is easily justified by respect for the material and by a deep aversion to seeing it treated as waste. The act itself of reclaiming wood from urban trees becomes a unique part of the woodworking process, and another dimension to the artistry of the results. Using reclaimed wood is not making something from nothing but is making something of value from a material that was on its way to becoming nothing.

As I related in the last chapter, the opportunity that I had to make a trestle table out of a 500-year-old bur oak tree from the Gatch family farm trumped every other consideration. Above all, I wanted to make that table from that wood. What I was paid covered my out-of-pocket expenses and none of my labor. But then I did not do it for the money.

The six main tasks are organizing a group of harvesting urban timber (HUT) volunteers, identifying sources of trees, contacting sawyers, selecting sawing sites, choosing the sawing method and lumber dimensions, and publicizing the effort. Below are basic ways to get these tasks done, and points to consider along the way.

Organizing a HUT Group

First of all, likeminded woodworkers can be found through local woodworking and wood turning clubs, lumber and woodworking stores (in-store billboard notices, handouts, as well as web site announcements and email to local customers), vocational departments in local schools and colleges, and from local saw mill owners. Each person contacted might then suggest at least one other person whom he or she knows who might also be interested.

At the first meeting, the goal of the effort should be clearly stated and accepted by everyone involved. Since woodworking cuts across occupations, gender, race and age, members of this group will probably have a variety of experiences, skills and training, which, properly harnessed, will contribute directly to the success of the effort.

Everyone involved should agree at the beginning on how labor, expenses and lumber should be shared. Whatever the formula, all must feel that it is fair and directly related to each person's contribution. Few problems can sink an effort faster, especially a voluntary one such as this, than the feeling that contributions and rewards are not fairly divided.

One person could serve as a treasurer, collecting enough from each participant to get the HUT effort underway. For example, you might assume that the first project(s) will yield 1,500 board feet: 1,200 board feet divided equally among participants and 300 either donated or swapped for logs. Assuming an out-of-pocket expense of 40 cents per board foot (25 cents per board foot for sawing and the other 15 cents for hauling and other expenses), the total cost is $600. A dozen participants would contribute $50 each and receive 100 board feet in return. Fifty cents per board foot for wood such as acacia, cherry, madrone, quarter sawn oak, and walnut—even undried and ungraded—is still a very good price. Consider providing extra wood to anyone who owns a truck and is willing to haul lumber to participants' homes.

Early on, everyone should agree in principle to donating some lumber to a public project or to an institution, such as a local vocational school. Donations of memorial benches and other public furniture made by participants, similar to those made for public parks in Lompoc, help establish this effort as a contribution to the entire community.

Also at the outset, a list should be compiled of the trees in your urban area, from the most desired to the least desirable. The local urban forester can help with this, as can tree services. Such a list serves as a guide for directing your group's limited resources to the most valued lumber and trees. In my area, walnut, cherry, oak, hard maple and ash are high on the list while beech, birch, cottonwood, pin oak, gum and soft maple are lower priorities. Unusual species should be considered as well: pecan, osage orange, and fruitwoods such as apple and pear.

This list should identify minimum log sizes. The best are generally 16 inches or larger in diameter at the small end of a log that is 10 feet or longer in length. Those 14 inches or larger in diameter at the small end of a log 8 feet or longer are judged as good. Logs 12 inches or larger in diameter at the small end of logs 8 feet or longer are considered fair. These dimensions and ratings should be adjusted downward for unusual but desirable trees: for example, fruit trees do not grow to the same diameters and lengths as oak or walnut, yet they are worth reclaiming. Osage can reach 16 inches and more in diameter but, because its trunks are often short—8 and 10 foot lengths (with no sweep) are rare. Even straight 6-foot lengths are uncommon.

Tentative decisions should be made about the way the logs are to be sawn (plain, rift, or quarter sawn) and about board thicknesses. Group members with particular projects can make their preferences known at this point.

Contact your state's department of forestry, natural resources, and the department that oversees solid-waste management for technical information, and ask about possible financial support for your local effort. Waste-management officials can explain your state's laws and any pending legislation on solid-waste disposal, and the implications for green waste and trees. Those of you in Indiana, New Jersey, Minnesota, and especially California, should expect very knowledgeable support. There is at least one university in nearly every state that has a forestry department that can provide technical and research support as well. Knowing that your HUT group exists allows state agencies, local agencies, and academic departments to keep you informed on legislation, possible grants and other forms of support. They can also keep you current on diseases and pests infecting trees and tell you what is being done in response. In some cases, you will not want to reclaim infected trees if doing so spreads the problem.

Identifying Sources of Trees

Next, contacts should be made with local parks departments, urban foresters, tree service companies, and local utility companies to determine what kinds of trees (species, size, and condition) they remove and where they dispose of them. Cemeteries, golf courses, and fruit tree orchards are occasional sources as well. So are housing and commercial land developers, although gathering and sawing a large number of trees from a major development project could be too much for a newly assembled group to tackle. Initial conversations should reveal how willing these sources are to cooperate. Do not be too surprised or disappointed if most are not initially interested (they may be later, as the idea catches on).

Parks departments and the local governments represented by urban foresters might have to check with senior administrators and city attorneys before they can agree to give up public trees. These conversations will go more smoothly if, from the outset, you express a willingness to pay for the logs, share the lumber or, at the very least, offer to pay an amount equal to hauling and tipping expenses. These trees are public property and their removal is a public expense, so you should not expect to get either for free.

If they have not dealt with this issue before, you might have to wait for an answer from utility companies until they decide whether it is a matter of company policy to cooperate. Again, offering to pay might help, while offering to split the lumber probably will not. As an added incentive, you could offer to make a donation to local schools, on their behalf, of lumber from trees they have taken down.

Tree service companies are likely to be the most flexible and quicker to decide whether to help. They are much more likely to be interested if you offer to pay for the extra effort required to load and haul leave logs in usable lengths (6 to 10 feet). In general, any offer you make to reduce expenses or to increase revenue will help your case. These companies are unlikely to be interested in receiving lumber instead of reimbursement. Recall from Chapter One that Lynn Erickson, a former log buyer for Minnesota Hardwoods, provided simple instructions on how the trees should be bucked into saw logs. With the tree services, Lynn found that informal discussions and the occasional presentation worked. By his estimate, prior to getting instructions, only about 10 percent of the logs were usable. However, once the companies learned how to cut, Lynn said that 90 percent were usable.

A list of your group's most desired species should be included. The same instructions and list should be provided to the parks departments and other participants as well.

Spray-paint the logs you expect to receive, and keep track of where they came from. The identity of the source will be useful later in describing the provenance of the pieces made from each tree. The specific origin of the reclaimed wood can add both sentimental and monetary value to the final products, especially furniture and crafts.

Finding Sawyers

At about the same time, several saw mill owners should be identified and contacted. Mill manufacturers can provide the names of owners in your area. Sawyers should be selected according to their rates, skill and experience, availability, mobility, and whether they can store and saw logs on their own property. I describe what to look for in pricing and skill in Chapter Three: recall that you want a sawyer who can saw for grade, that is, one who knows how to saw the best quality boards from each log. You may want one who can quarter-saw as well, especially if you are going to get oak and sycamore logs. Best of all, perhaps one or more mill owners could be persuaded to join your HUT group. You may have several offers from owners of chain saw, circular saw, and thin-kerf band saw mills. The advantages and disadvantages of each are also discussed in Chapter Three.

Selecting Sawing Sites

Next, select and prepare a sawing site. If there is sufficient space, the parks department might allow the mill on-site, especially if you are sharing the lumber with the department. Advance preparations, such as lining the logs up so that they can be easily loaded onto the mill, allows the efficient use of the sawyer's time. A mobile sawyer is needed here, as is a way of securing the sawmill overnight if the work takes more than a day. Very likely, street trees and those on private property will have to be hauled to the sawyer or to another place for storage. A mill owner with enough property to accommodate the logs offers an important advantage over hauling to a secondary storage and sawing site. For a fee, perhaps the tree service company will allow you to use its property for short-term storage and sawing.

Sawing Choices

Firm decisions about how the logs are to be sawn—method and dimensions—should have been made by the time the logs and the sawmill are brought together. Basic methods of sawing for grade are discussed in Chapter Three. On sawing days, participants should be present to stack and load the lumber as it comes off the mill. Assuming that the lumber is divided among participants, everyone should sticker and end-coat their own lumber as soon as possible. Participants with trucks will make hauling lumber a lot easier.

Publicizing the Effort

Now, after several projects have been successfully completed and most everyone involved feels confident about the process, suggest to local newspapers and television stations that they do a story on the effort.

Favorable publicity is free advertising for participating tree-service companies, and it informs the public that local parks departments are trying to make the best use of public funds and property. Publicity, especially if it involves donations of wood or the use of special trees to make public furniture, will also expand your sources and might change the minds of those who had once been reluctant to sign on. It should also bring in additional volunteers, including some who are more interested in contributing to community improvement projects than they are in woodworking. Newspaper articles serve as a future reference for property owners: readers will cut and save such articles for some time. I still receive calls from people who held onto local newspaper articles describing my own effort from the mid-nineties.

In the reporting, plainly state that your HUT group does not take down trees in exchange for the wood, otherwise, you will receive numerous calls from homeowners offering trees for free if you will just do the removal for free. I refer all of these callers to a local tree-service company. However, knowing that lumber is donated to public institutions, and projects are made for public use, is incentive for some to donate trees taken down at their own expense.

Assessing Tree Quality and Size

In the initial effort, you can rely on professionals working for the parks department, utility companies, and tree service companies, to provide accurate information on the quality, species and size of the trees being harvested. After your effort receives local publicity, you will receive offers from home and property owners. Someone will have to personally examine the trees being offered to determine whether in terms of quality, species, and size they are worth your group's effort to reclaim them for lumber. The basic quality issues are the presence of defects, including embedded objects the tree did not grow, and curvature (sweep) of the tree trunk. As to species, owners often do not know what kind of tree they are offering, or they identify it incorrectly.

Accurately sizing the tree is very important. From my experience, a surprising number of people seem not to know the difference between diameter and circumference. Many describe their trees as large: trunks "over 3 feet" is the size I hear most frequently. If the diameter is more than 3 feet, the tree is large. If the circumference is about 3 feet, then the tree has a diameter of roughly 1 foot and is not so large after all (the circumference of a circle equals pi, 3.14, multiplied by its diameter—therefore the diameter of a circle equals its circumference divided by 3.14). The difference in board feet is substantial. A straight, unflawed 16-foot log with a 40-inch diameter contains just over 1,200 board feet (by the International 1/4-Inch rule). By contrast, a log the same length, but with a 40 inch circumference, would be about 13 inches in diameter and would contain (by the same rule) about 115 board feet, less than one-tenth of the amount in the larger log.

Safety is Always the First Priority

Review the safety recommendations in Chapter Two on felling, limbing and bucking trees if you decide to do some of this work yourself. Also, you should not fell a tree alone, when impaired by alcohol or medications, or when the tree itself poses special hazards (for example, when it is leaning, caught in another tree, cabled, or lying on a building). If the tree is down and needs limbing and bucking, everyone involved should review the rules for safe chain saw use. Wear protective gear, make sure the saw is well-maintained, start it properly, and do not overextend yourself while using it. Exercise care when limbing, being especially mindful of limbs under pressure (for example, spring poles) from the tree trunk itself. Watch for a trunk to shift as limbs beneath it are cut.

Also review the means of skidding and hauling logs outlined in Chapter Three. Do not try lifting heavy logs on your own. Roll or carry them across yards and driveways using dollies or sheets of plywood to minimize damage to yourselves and to lawns. Having the property owner sign a release, and photographing the area before starting, demonstrates to the property owner that your group is thinking ahead about any questions that might arise regarding damage.

A winch attached to a truck can expedite the removal of logs; however, remember that the cable is under stress when in use, and take care to minimize the possibility of injury should it break (for example, keeping the area clear and draping a blanket over the cable to prevent it from whipping about should it, or an attachment, fail). Once loaded into a truck or onto a trailer, logs and lumber should be tied down securely.

Next Steps

As basements and garages fill up, your HUT group will have accumulated enough air drying lumber to meet your own demands. At this point, a decision must be made about what to do next. One option is to continue at the same level, acquiring as much as the group can absorb and passing on the rest of the available trees. Another option is to enlarge the scale of the existing effort to accommodate the growing supply of logs and lumber. The extra can be stockpiled (requiring a storage area) or sold. As long as there is space most logs, properly stored, can be held for a few years until the lumber is needed.

Selling Logs and Lumber

Selling the logs, or sawing and selling the lumber, raises three big questions: who are the buyers, how will the logs and lumber be delivered, and what kind of organization will be required? To start with, where do you find buyers?

1. Of course, amateur woodworkers, including turners and carvers, who happen not to be participating in your HUT effort will be interested. They can be contacted the same way participants were; through posted notices, email advertisements, and word-of-mouth at woodworking stores. You could offer to conduct a log-sawing demonstration at one of these stores and then give away small samples of wood at the demonstration. The same approach can be taken with local and regional woodworking clubs.
2. Custom furniture craftsmen and cabinetmakers may be interested. Unless they are building their inventories of lumber and are, therefore, willing to buy air-dried wood, they may require kiln-dried lumber that can be used immediately. They may be interested in unusual wood, fruitwoods for example, crotchwood, and lumber with flaws that can be used as design elements. Companies that are SmartWood^{cm} certified may be especially interested (http://www.brandsystems.net/smartwood/). You can suggest that potential customers for these businesses are owners of trees who want something special made from the wood.
3. Architectural millwork firms are potential buyers. You will have to saw to their specifications and have the species available they use. Unless they are stockpiling, they, too, may require kiln-dried wood. Many of these firms are members of the Architectural Woodwork Institute, which publishes quality standards for members. Getting a copy (cited in the References under Architectural Woodwork Institute) and reviewing the standards might help. Again, companies that are SmartWood^{cm} certified may be more receptive.
4. Groups such as the Amish, who manufacture wood furniture from oak and cherry, are potential buyers of fairly large quantities. They could be in the market for both logs and lumber.
5. While sawmill and kiln operators will not buy urban logs, they might be willing buyers of the lumber, since the risk of their having to saw urban trees has been effectively eliminated.
6. Independent lumber retail stores may be willing to either buy your wood or take it on consignment. Some, like EcoTimber® in Berkeley, California, specialize in selling reclaimed and recycled wood and lumber sawn from sustainably managed forests and wood lots. There are other companies in California and across the country that have been certified under

SmartWood^{cm} Rediscovered Wood Program that could be especially interested.

7. Look for businesses that manufacture wood products, such as plaques, kitchen cutting boards and implements, decorative boxes (including jewelry boxes and humidors), picture frames, shelving, wooden toys, and dowels. I found a company in my area making dowels and spindles that was willing to buy wet hardwoods for prices posted in the Hardwood Market Report. They graded and dried their own wood and were pleased to have a potential local supply cut to their specifications.
8. Lower quality wood could be sold to manufacturers of pallets, skids, and cable reels. The list of companies by state who are members of the National Wooden Pallet and Container Association can be found on the Association web site: (http://www.-nwpca.com/ search state.htm).
9. Thick planks of lower quality wood for replacement trailer decking might be of interest to local trucking and construction companies.
10. Stone cutting companies place timbers just beneath the blocks of stone being cut to protect the blade and the concrete piers on which the material rests from being damaged. There is a constant demand for new timbers because the ones in use are cut so many times that they must be replaced frequently.
11. Landscaping firms can also make use of lesser but durable wood as landscaping ties, posts, borders for playground areas, fencing, hillside steps, and railings. Using untreated wood for play areas should be especially appealing to those who are concerned about the possible adverse health effects of wood preservatives on children.
12. If your group is not already working with them, local parks departments may be interested in landscaping ties and lumber for outdoor furniture. If you are working with them, one way to sell extra wood is to place notices in shelter houses and on park web sites. Also, if the department sells firewood to the public, stack the lumber near the firewood. Without any extra effort, you may sell all the lumber you have to offer.
13. Schools that offer woodworking courses may be interested in purchasing lumber. If your group is already donating some, they may be interested in buying more.
14. If there are any veneer manufacturers in your urban area, you might contact them to see if they would be interested in purchasing your very best logs. With two possible exceptions, this could be a difficult sell. I explored this possibility with one manufacturer and found that there is concern about imbedded metal that could seriously damage the veneer knives. Large companies deal in logs by the tractor-trailer load and by practice are not inclined to spend their buyers' time inspecting just a few logs. I argued that there is no guarantee that forest logs are free of foreign material and that good visual inspection could catch the urban logs with problems, especially after they have been debarked. Also, since they spend most

of their work week driving hundred of miles to inspect logs, why not spend a little extra time, tacked on to the end of the lunch hour, to look at a few local logs. While not totally dismissed, I received a cordial but cool response to the idea. There are two exceptions that might be more promising: companies may be more willing to inspect sources of highly-figured wood such as burls (from ash to eucalyptus to walnut) and bird's eye, curly and quilted maple (both of which you might want to consider keeping for your HUT group). Second, veneer companies that have received the SmartWood^{cm} certification might be more flexible about purchasing urban logs.

15. You can try wholesale lumber dealers who are willing to buy wet lumber by the truckload at hardwood market prices. I found one that was willing to negotiate for 10,000 board feet of quarter-sawn red oak. Professional grading is very likely to be a condition of the sale.

Once the sales are made, the material must be delivered. Members of your HUT group can use their own, or rental, trucks to deliver small amounts of lumber. Large quantities of logs and lumber require forklifts, or trucks equipped with hoists for loading, and large trucks or tractor-trailers for hauling. You will have to hire the operator and equipment. However, sod companies are another possibility. A sawyer friend of mine is the retired owner of a sod farm and has access to heavy-duty forklifts as well as large trucks and tracker-trailers. For a fee, he has hauled lumber and picked up logs for me (this works better in the cold and dry seasons when sod sales drop off).

Once you begin acquiring and selling logs and lumber, you should consider incorporating your group as a tax-exempt organization. Filing for incorporation must be done through both your state (in Ohio, it is through the office of the Secretary of State) and the Internal Revenue Service. Each state has its own forms and procedures. For the IRS, to start with, see publication 557, *Tax-Exempt Status for Your Organization* (available for downloading at http://www.irs.gov/pub/irs-pdf/p557.pdf). As a tax-exempt organizaton your HUT group is allowed to sell logs and lumber, pay reasonable wages, and purchase equipment. As a first purchase, your group should consider buying its own mill and having several members who can competently operate it. Nonprofit status also allows your group to accept both governmental and private grants for your work.[1] Perhaps one of your HUT members will be a tax attorney or an accountant who can provide advice on the best choice among the 501(c) options.

Having a larger formal organization with experience and equipment could put your group in the position of taking on the challenge of reclaiming trees from land that is being cleared for residential or commercial development, infrastructure projects such as highway expansion, or orchards that are being cleared. You could acquire enough lumber here to make a commercial sale.

As your group grows, more formal arrangements to share lumber could be made with local parks departments. Your group's collaboration with public officials, and the favorable publicity attached, would encourage them to reclaim as many trees as they can. If you are very fortunate, you may find women such as Cindy McCall or Judy O'Bannon in positions to promote the idea and help make it happen (there are probably a few men who could do the same). A similar arrangement with the local utility companies could encourage them to reclaim the trees they cut. As these relationships grow, perhaps one or both would be willing to let your HUT group use their vacant property for temporary log storage and as cutting sites.

Consider working closely with one or more tree services in a cooperative arrangement. If they have the property, and are willing to fell trees as saw logs, then your group can saw and sell the lumber and share the proceeds. If this works well, one or more of these companies, like West Coast Arborists and East-West Urban Forest Products, might launch a new business of reclaiming trees and selling the lumber. Your group can help demonstrate that there can be profit in this commitment.

A Final Thought

The laws that govern the physical world operate in spite of what we want—this is what makes physics so annoying at times (the slice of bread that lands jelly-side down, the coffee that somehow spills into the computer keyboard, or the kickback from a table saw). However, the Laws of Demand and Supply operate because of what we want; indeed, demand in particular is based on our preferences, what we want and what we are willing to pay for it. If we want to make reclaiming three to four billion board feet of wood wasted every year into a project that will work, then it will. If, as consumers and urban residents, we favor reclaimed wood and wood products, if we make this an important attribute of the wood products we buy, then this will encourage businesses, government, and community groups to meet that demand.

Trees provide so much: they are sources of shelter, food, medicine, habitat, oxygen, and artistic material. In our cities, they clean the air we persistently foul, muffle the noise we constantly make, and hide the things we do not want to see. And, when we stop long enough to simply look at them, they can provide a moment of serenity during a busy day on a crowded city street. There is no good reason to discard this gift once the trees come down. We have such a knack for turning so many natural things of value into waste. By reclaiming urban trees, we preserve and extend the value of one of nature's greatest gifts to us. To do otherwise is both a material and a spiritual loss.

Endnotes

Introduction

1. Telephone communication with Ryan L. Short, U.S. Bureau of the Census, August 7, 2001. The basic reference for these definitions is: Bureau of the Census. November, 1994. *Geographic Areas Reference Manual,* U.S. Department of Commerce, Washington, DC, chapters 9 and 12. Available at http://www.census.gov/geo/www/garm.html

2. Except in six New England states where metropolitan areas are defined by cities and towns, not counties.

3. For OMB definitions and listing of metropolitan areas, go to http://www.whitehouse.gov/omb/bulletins/95-04attachintro.html

4. Dwyer, John F., David J. Nowak, Mary Heather Noble, and Susan M. Sisinni, August, 2000. *Connecting People With Ecosystems in the 21st Century: Ann Assessment of Our Nation's Urban Forests,* U. S. Department of Agriculture, Forest Service, Northeastern Research Station, Syracuse, NY: 27.

5. Bratkovich, Stephen M. October, 2001.*Utilizing Municipal Trees: Ideas from Across the Country,* NA-TP-06-01, U. S. Department of Agriculture, Forest Service, Northeastern Area, St. Paul, MN: 1 - 3.

6. James L. Howard. April, 2001. *U.S. Timber Production, Trade, Consumption, and Price Statistics 1965-1999,* Research Paper FPL-RP-595, U.S. Department of Agriculture, Forest Service, Forest Products Laboratory, Madison, WI, Table 12: 33.

7. The most recent data are for 1997. Logging passed commercial fishers as the nation's most dangerous occupation. At 128 deaths per 100,000 workers, the fatality rate for loggers is 27 times the average for all occupations. For further information, see the Eric F. Sygnatur. winter, 1998. "Logging is Perilous Work", *Compensation and Working Conditions,* Bureau of Labor Statistics, Washington, DC, p. 3 - 9. This report is available on the Bureau of Labor Statistics web site at http://stats.bls.gov/iif/oshwc/cfar0027.pdf. Falling trees and limbs cause 65 percent of the fatalities. An additional 5 percent are caused by rolling logs. Arborists and tree services, not commercial loggers, fell trees in the more densely populated parts of urban and metropolitan areas. They are not injured or killed by falling trees and limbs in the same proportion as loggers because felling trees is a small part of what they do. Even so, in 1997 about 38 percent of the fatalities among workers in this industry arose from being struck by something, most likely tree trunks and limbs. A third died from falls (see http://stats.bls.gov/iif/oshcfoi1.htm). Whether viewed as logging or as tree care, felling urban trees is dangerous, especially for the untrained.

Chapter One

1. Anonymous. November, 2000. "The State of Garbage: 12th Annual *BioCycle* Nationwide Survey", BioCycle: 40 - 48.

2. Environmental Protection Agency. September, 1999. *Characterization of Municipal Solid Waste in the United States: 1998 Update*, EPA 530-R-99-021, Washington, DC.

3. Whittier, Jack, Denise Ruse, Scott Haase. 1994. *Final Report Urban Tree Residues: Results of the First National Inventory*, ISA Research Trust, Savoy, IL.

4. Anonymous. November, 2000. "The State of Garbage: 12th Annual BioCycle Nationwide Survey", *BioCycle*: 40 - 48.

5. Telephone communication with Edward Lempicki, June 14, 2001.

6. Telephone communication with Ed Cesa, June 13, 2001.

7. Royce, Rebecca. December, 1995. "Program aims to turn 'waste' trees into source of profitable sawlogs", *Woodshop News*: T37 - T39.

8. Lempicki, Edward, Ed Cesa, and J. Howard Knots. June, 1994. *Recycling Municipal Trees: A Guide to Marketing Sawlogs from Street Tree Removals in Municipalities*, USDA Forest Service, Washington, DC.

9. Telephone communication with Edward Lempicki, June 14, 2001.

10. Blanche, C. E. and H. F. Carino. October, 1996. "Sawmilling Urban Waste Logs: An Income-Generating Option for Arborists", *Arborist News*: 45 - 48.

11. Benvie, Sam. 2000. *The Encyclopedia of North American Trees*, Firefly Books, Ltd., Buffalo, NY: 268.

12. Smith-Fiola, Deborah. 2000. "Integrated Pest Management", Kuser, John (ed.). 2000, *Handbook of Urban and Community Forestry in the Northeast*, Kluwer Academic/Plenum Publishers, New York: 261 - 286.

13. Caldwell, Brian. August, 2001. "Program restores elms to nation's main streets", *Woodshop News*, XV (9): 6. In addition, researchers at the University of Abertay in Dundee, Scotland have discovered a promising way to genetically modify the elm to resist Dutch elm disease. See Peter Gorner. August 30, 2001. "Science crafts hardier elm", *Chicago Tribune*.

14. Minnesota Department of Natural Resources. June, 1994. *Final Report: Urban Tree Utilization Project*, St. Paul, MN.

15. Fisher, James A. March, 1992. *Urban Tree Residue: An Assessment of Wood Residue from Tree Removal and Trimming Operation in the Seven-County Metro Area of Minnesota*, Division of Forestry, Minnesota Department of Natural Resources.

16. Telephone communication with Phil Vieth, June 20, 2001.

17. Telephone communication with Frank Kilibarda, June 15, 2001.

18. Telephone communication with Phil Vieth, June 6, 2001.

19. Telephone communication with Lynn Erickson, June 25, 2001.

20. Minnesota Department of Natural Resources. June, 1994. Final Report: *Urban Tree Utilization Project*, St. Paul, MN.

21. Telephone communication with Dennis Ludivig, June 21, 2001.

22. www.ciwmb.ca.gov/2000Plus/History.htm

23.www.ciwmb.ca.gov/lgcentral/Rates/Diversion/RateTabl.htm

24. Telephone communication with Eric Oldar, June 25, 2001.

25. I believe everyone should take a plane or helicopter ride over their own and surrounding homes and neighborhoods. There is no better or quicker way to acquire a sense of the size and extent of the urban forests we live in than to see them from above. If you cannot take the flight, the alternative is to see your neighborhood and city from satellite by going to http://terraserver.homeadvisor.msn.com/default.asp.

26. Nakashima, George. 1988. *The Soul of a Tree: A Woodworker's Reflections*, Kodansha International, Tokyo.

Chapter Two

1. As required by the 1974 Forest and Rangeland Renewable Resources Planning Act (RPA) the Forest Inventory and Analysis (FIA) program has been collecting information on American (non-urban) forests for over 60 years. Like the population census, the FIA Program collects and interprets data on the size, composition, distribution, health, as well as current and projected trends in the nation's forests. For a brief summary of the program go to http://www.fia.fs.fed.us

2. Gregory McPherson and others recently conducted a comprehensive urban tree study in Sacramento, California. Though the composition of this city's forest has changed over the years, public involvement in supporting, protecting, and helping to sustain the forest has been fairly constant. The details of the history of Sacramento's urban forest are described in "From Nature to Nurture: The History of Sacramento's Urban Forest", *Journal of Arboriculture*, March, 1998, 24(2): 72 - 88. A sample survey of residential lots in Sacramento was conducted to calculate the number, size, species, and condition of residential trees. Owners were interviewed about the numbers of trees they planted, their reasons, and their maintenance efforts. Resident actions and attitudes are critical since

about three-fourths of all of Sacramento's trees are on residential property. Joshua Summit and E. Gregory McPherson, "Residential Tree Planting and Care: A Study of Attitudes and Behavior in Sacramento, California", *Journal of Arboriculture*, 24(2): 89 - 111. Survey data allowed an estimate to be made of the number of trees in Sacramento—about 6 million—as well as an assessment of the health and sustainability of the city's forest. The results were encouraging: about 70 percent of the trees were healthy and they were well distributed by age and species. The most abundant species were well suited to growing conditions in Sacramento. "Structure and Sustainability of Sacramento's Urban Forest", *Journal of Arboriculture*, July, 1998, 24(4): 174 - 190. Now this is the kind of study that cities with a serious interest in their urban forests should conduct. Knowing the history of the forest, especially the role of residents both as property owners and tree care volunteers, is a way to gauge the commitment urban residents are willing to make. These studies were not conducted to determine the volume of saw logs that Sacramento's urban forest might yield. However, the data on the inventory, tree species, size, condition, planting and replacement could be used to make estimates of the log volume (assuming tree removal was conducted with saw logs in mind, not just expeditious disposal). In general, calculated for the major species, volume multiplied by prices gives an estimate of the total revenue that the sale of logs and lumber might generate. Comprehensive urban forest studies have also been conducted in Chicago, Illinois and Oakland, California. For the Chicago study, see McPherson, E. G., D. J. Nowak, R. A. Rowntree, eds., *Chicago's urban forest ecosystem; results of the Chicago urban forest climate project*, Gen. Tech. Rep. NE-186. Radnor, PA, U.S. Department of Agriculture, Forest Service, Northeastern Forest Experiment Station: 3 - 18, 140 - 164. For the Oakland study, see David J. Nowak, *Urban forest development and structure: analysis of Oakland, California*, Ph.D. Dissertation, University of California, Berkeley, 1991. These two reports are summarized and compared in Dwyer, John F., David J. Nowak, Mary Heather Noble, and Susan M. Sisinni, August, 2000. *Connecting People With Ecosystems in the 21st Century: An Assessment of Our Nation's Urban Forests*, U. S. Department of Agriculture, Forest Service, Northeastern Research Station, Syracuse, NY: 47 - 60.

3. The Forest Service and the EROS Data Center are currently developing higher-resolution urban maps—down to 30 meters from current 1000 meters (1 kilometer). The U.S. Geological Survey (USGS) operates EROS, the Earth Resources Observation Systems Data Center.

4. Dwyer, John F., David J. Nowak, Mary Heather Noble, and Susan M. Sisinni, August, 2000. *Connecting People With Ecosystems in the 21st Century: An Assessment of Our Nation's Urban Forests*, U. S. Department of Agriculture, Forest Service, Northeastern Research Station, Syracuse, NY. Methods used to estimate tree coverage and tree populations are described on pages 108 - 112. Tree coverage estimates for the 48 contiguous states are available in this publication for all counties, urban places, places within urban areas, and metropolitan areas. Tree population estimates are available at the metropolitan level by states as listed in Table 2.1.

5. Since most urbanized areas and urban places are located in the larger metropolitan centers, nearly all of the 3.8 billion urban trees are included in the larger count of 74.4 billion. Trees in those few urban areas and places outside metropolitan centers are not included in the metro count. If they were added, the estimate would go above 74.4 billion. Since the Forest Service did not merge the two estimates, the metro count is slightly conservative. E-mail communication with David J. Nowak. August 3, 2001.

6. Telephone communication with David J. Nowak, July 30, 2001.

7. www.nps.gov/seki/bigtrees.htm

8. Hall, Carl T. August 23, 1998. "Staying Alive High in California's White Mountains grows the oldest living creature ever found", *San Francisco Chronicle*. Another tree named Prometheus, estimated to be at least 4,900 years old, was cut down in 1964 by a geography professor from the University of Utah with the help of the USDA Forest Service.

9. Wilson, Brayton F. 1984. *The Growing Tree* (revised edition), The University of Massachusetts Press, Amherst, MA: 14 - 15.

10. Kasindorf, Martin. 2001. "Trees take root as energy savers", www.usatoday.com/news/nation/2001/07/24/tree.htm

11. Nowak, David J. and John F. Dwyer. 2000. "Understanding the Benefits and Costs of Urban Forest Ecosystems", *Handbook of Urban and Community Forestry in the Northeast*, Kluwer Academic/Plenum Publishers, New York: 12 -13, 16. The combination of transpirational cooling, alteration of wind speed and direction of flow, and blockage of radiation is how trees produce both cooling and warming effects. Given that they are large enough, whether they have a cooling effect in hot weather and a warming effect in cold weather depends on the type of trees and where they are located relative to homes and commercial buildings. For example, in warm climates, conifers south and southwest of buildings will have a greater cooling effect than similarly placed deciduous trees. However, deciduous trees lose their leaves in the winter allowing sunlight through to warm buildings in climates where warm and cold weather alternate. In the latter climate, one or (even better) two rows of conifers arranged in a U or L shape facing the prevailing winter winds can reduce heating costs.

While appropriate placement reduces heating and air conditioning costs, poor choice of species and their placement can have the opposite effects.

12. Nowak and Dwyer, 13 - 14. When a tree dies and decomposes, sequestered CO_2 returns to the atmosphere unless the tree is sawn into lumber. Sawing trees into lumber, instead of letting them decompose in landfills and on-site, helps reduce CO_2, a major greenhouse gas that contributes to global warming. This is one of many advantages of harvesting urban trees. See David Novak. September 11 - 15, 1999. "Impact of Urban Forest Management on Air Pollution and Greenhouse Gases", *Proceedings of the Society of American Foresters, 1999 National Convention*, Portland, OR: 143 -148. To be strictly accurate, we would have to add CO_2 emitted by the gas engines of sawmills and chain saws used to cut the trees to the amount of CO_2 reduced by trees to obtain the net reduction. As long as the net reduction is greater than zero, trees and the lumber that comes from them are making their contributions to cleaner air. David Nowak and Daniel Crane have estimated that at present the urban forests of the lower 48 states are holding 700 million tons of carbon. As they point out, we can take advantage of this as long as the machinery we use to maintain and fell the trees does not produce more CO_2 than the trees themselves are sequestering. For a detailed description of how urban forests can reduce CO_2 and how the amount sequestered is calculated, see E. Gregory McPherson and James R. Simpson. January, 1999. *Carbon Dioxide Reduction Through Urban Forestry: Guidelines for Professional and Volunteer Tree Planters*, Gen. Tech. Rep. PSW-GTR-171, Albany, CA: Pacific Southwest Research Station, Department of Agriculture, Forest Service. In addition to heating, cooling and cleaning urban air, urban trees also control storm water runoff, a growing problem for cities increasingly covered by asphalt and concrete. The nonprofit conservation organization, American Forests, has developed the Regional Ecosystem Analysis (REA), an analytical process that calculates the economic values of the air and water effects urban trees have in specific urban areas. The process utilizes satellite imagery, on-ground field surveys, and American Forest's own Windows-based GIS program, CITYgreen. See http://www.americanforests.org/ productsandpubs/citygreen/cg5.php.

13. Haygreen, John G., Jim L. Bowyer. 1996. *Forest Products and Wood Science* (3rd ed.), Iowa State University Press, Ames, IA: 58.

14. Haygreen and Bowyer, 74.

15. Forest Products Laboratory. 1999. *Wood Handbook, Wood as an Engineering Material*, General Technical Report, FPL-GTR-113. U.S. Department of Agriculture, Forest Service, Forest Products Laboratory, Madison, WI

16. Hoadley, R. Bruce. 2000. *Understanding Wood: A Craftsman's Guide to Wood Technology*, Taunton Press. Wilson, Brayton F. 1984. *The Growing Tree* (revised edition), The University of Massachusetts Press, Amherst, MA.

17. Wilson, Brayton F. 1984. *The Growing Tree* (revised edition), The University of Massachusetts Press, Amherst, MA.

18. Keator, Glenn. 1998. *The Life of an Oak*. Heyday Books and the California Oak Foundation. Berkeley, CA.

19. Lempicki, Edward, Ed Cesa. 2000. "Recycling Urban Tree Removals" in Kuser, John, ed., *Handbook of Urban and Community Forestry in the Northeast*, Kluwer Academic/Plenum Publishers, New York: 362.

20. Plotnik, Arthur. 2000. *The Urban Tree Book: An Uncommon Field Guide for City and Town*, Three Rivers Press, New York: 127 - 128.

21. Forest Products Society. 1999. *Wood Handbook, Wood as an Engineering Material*, Madison, WI

22. ForestWorld, *Woods of the World* (version 2.5), CD-ROM, Middlebury, VT. This and other wood-related information are accessible at www.forest-world.com.

23. Forest Products Laboratory. USDA Forest Service.1999. *Wood Handbook, Wood as an Engineering Material*, General Technical Report, FPL-GTR-113, U.S. Department of Agriculture, Forest Service, Forest Products Laboratory, Madison, WI: 4-31 - 32.

24. Whittier, Jack, Denise Ruse, Scott Haase. 1994. *Final Report Urban Tree Residues: Results of the First National Inventory*, ISA Research Trust, Savoy, IL.

25. www.dfr.state.nc.us/storm

26. www.weyerhaeuser.com

27. Two web sites that provide extensive information on the beetle are the USDA Forest Service's www.na.fs.fed.us/spfo/alb/index.htm and the University of Vermont's Entomology Research Laboratory at www.uvm.edu/albeetle/index.html.

28. www.fs.fed.us/foresthealth/sod_ca.html and http://www.cnr.berkeley.edu/garbelotto/english/sod_diagnostic_report_final.pdf

29. Leininger, Theodor, David L. Schmoldt, and Frank H. Turner. June 18, 2001. "Using Ultrasound to Detect Defects in Trees: Current Knowledge and Future Needs", *First International Precision Forestry Symposium*, University of Washington, Seattle, WA. Also see, Xu, Zicai, Theodor D. Leininger, Andy W. C. Lee, Frank H. Turner. 2001. "Chemical Properties Associated with Bacterial Wetwood in Red Oaks", *Wood and Fiber Science*, 33(1): 76 - 83.

30. Hall, Guy H. 1998. *The Management, Manufacture and Marketing of California Black Oak, Pacific Madrone and Tanoak*, Western Hardwood Association, Camas, WA: 57.

31. Reported to me by John "Rusty" Dramm, Forest

Products Specialist, Forest Products Laboratory, USDA Forest Service, Madison, WI. November 2, 2001.

32. The Z133 Committee of the American National Standards Institute (ANSI) established safety standards for tree work in 1972. These standards not only cover tree trimming and felling but also how this work should be done in the vicinity of electric lines, the use of harnesses for aerial work, operation of cranes and hoists, various hand operated and powered cutting tools, cabling and lowering limbs, and limbing and bucking. See American National Standards Institute. October 19, 2000. *American National Standard for Arboricultural Operations - Pruning, Repairing, Maintaining, and Removing Trees, and Cutting Brush - Safety Requirements*, ANSI Z133.1-2000, New York. The standard was created as a result of the efforts of a New York woman whose son was killed in a tree-trimming accident.

33.http://www.natlarb.com/downloads/laws%20and%20stds%20OSHA3.pdf

34. There is a dark side to this practice. Apparently some residential developers will clear cut their property (a legal method of harvesting timber) and sell the logs as a way of avoiding local development codes that require residential green buffer zones. The developers clear first and then submit their development plans afterward, thus skirting the buffer requirement. At least one state, North Carolina, having caught onto this practice, passed legislation allowing some local authorities to deny a development application for clear-cut land up to five years after the land has been cleared. The developer can choose either to clear the land and not use it for residential development for up to five years, or can leave acceptable buffer zones and apply for immediate residential development. Arthur McLean. April 22, 2002. "Urban logging raises concerns", *Fayetteville Online, Fayetteville Observer*, http://www.fayettevillenc.com//obj_stories/2002/apr/n22cut.shtml

35. Mary Carmen Cupito and Gayle Harden-Renfro. July 27, 1982. "Tree trimmer loses arm, leg", *The Cincinnati Post*: 10C. Paul Furiga. July 27, 1982. "Man Pinned In Tree When Limb Crashes Down", *The Cincinnati Enquirer*: D1.

36. To get a more precise feeling for the dangers involved, go to the Forest Resources Association website (http://www.apulpa.org/) and under *Timber Harvesting Safety*, read the accident reports under *Safety Alerts*.

37. Homeowners have offered their trees to me for free if I will just take the trees down for free. Having found out what the price is for removal, I believe these offers are motivated by a desire to avoid or at least reduce what the homeowners will have to pay. While I understand the reason for such offers (there's certainly no harm in asking), I never make these deals, for the reasons I have explained. Instead, I recommend a reputable tree service. In response to some of these calls, I try to persuade owners to let healthy trees stand if they are not hazards but just a nuisance. More than any other species, I have defended walnut trees, usually without success.

38. See "Manual Timber Felling Hazard Recognition", Loss Control Overviews LCO14 under *Timber Harvesting Safety* at http://www.apulpa.org/ for the complete list that I have modified here to fit the urban setting. Also reflected in the list are recommendations from Occupational Safety and Health Service. *A Guide to Safety in Tree Felling and Cross Cutting*, February, 2001. Department of Labour, Wellington, New Zealand (available at http://www.osh.dol.govt.nz/order/catalogue/pdf/treefell.pdf).

39. See www.osha-slc.gov/SLTC/logging_advisor/mainpage.html

40. For more detailed information on bore cutting as well as felling, limbing, and bucking, see Forest Industry Safety and Training Alliance. April, 1999. *Loggers' Safety Training Guide*, Rhinelander, WI. Doing this cut properly and safely requires professional training. The procedures described here and in the section on safe chain saw use are based on the Game of Logging® chainsaw training method developed by Soren Erikson and Tim Ard. Designed for anyone who uses a chain saw—loggers, arborists or woodlot owners—this widely-used approach involves intensive training in safe chain saw use, reinforced by competitive demonstrations among students using what they have learned. Ultimately, students participate in nationwide contests where they can win monetary awards. Additional information on FISTA is available on their web site: http://hosting.newnorth.net/fista/.

41. Vondriska, George. August, 2000. "Bandsaw Resawing", *American Woodworker*, no. 81: 46 - 51.

42. Smiley, E. Thomas, Bruce R. Fraedrich, and Peter H. Fengler. 2000. "Hazard Tree Inspection, Evaluation, and Management", John Kuser (ed.), *Handbook of Urban and Community Forestry in the Northeast*, Kluwer Academic/Plenum Publishers, New York: 250.

43. Blenk, Stephen H. August, 1994. "Working with Burl", *American Woodworker*, 39: 48 - 51.

44. U.S. Consumer Product Safety Commission. no date. *Product Summary Report, All Products Injury Estimates for Calendar Year 2000*, National Electronic Injury Surveillance System, National Injury Information Clearinghouse, U.S. Consumer Product Safety Commission, Washington, DC. See product code #1411. The estimate for chain saws and all other product estimates are based on reports from a sample of hospital emergency rooms around the country. At a 95 percent confidence level, the actual number of chain-saw-related injuries treated in emergency rooms lies somewhere between 22,000 and 31,423. The CPSC states that emergency room records simply state that

the product or recreational activity was associated with the injury, not that one or the other actually caused the injury. I would say, however, that with chain saws, association and causation are very close, if not synonymous. This reporting system does not include injuries treated outside emergency rooms, for example, by EMS personnel on-site or by those who treat themselves.

45. Forest Industry Safety and Training Alliance. December, 1999. *Arborists' Chain Saw Safety Training Guide*, Rhinelander, WI. and Forest Industry Safety and Training Alliance. April, 1999. *Loggers' Safety Training Guide*, Rhinelander, WI.

46. A NIOSH review of the epidemiological evidence on HAV concluded that there is "strong evidence of a positive association between high level exposure to hand-arm vibration (HAV) and vascular symptoms of hand-arm vibration syndrome (HAVS)." Furthermore, increased intensity and duration of exposure directly raises the risk of developing HAV, shortens the time before the onset of symptoms, and intensifies the symptoms once they start. Bruce P. Bernard, (ed.). 1997. *Musculoskeletal Disorders (MSDs) and Workplace Factors: A Critical Review of Epidemiologic Evidence for Work-Related Musculoskeletal Disorder of the Neck, Upper Extremity, and Low Back*, U.S. Department of Health and Human Services, Centers for Disease Control and Prevention, National Institute for Occupational Safety and Health, Publication No. 97-141, Cincinnati, OH, Chapter 5c: 1 (available at http://www.cdc.gov/niosh/pdfs/97-141b.pdf). Additional information for field use is available at http://www.cdc.gov/niosh/elcosh/docs/d0200/d000259/d000259.pdf.

Chapter Three

1. Horse logging, the use of teams of draft horses or mules to pull logs from small and medium sized wood lots and forests, is a low-impact option. A major advantage of this method is that it is less intrusive and damaging to the forest than logging with machines. Because they do not weigh as much, horses will not compact the soil like felling and skidding machines, nor do they require a network of intrusive and damaging logging roads. Horse teams also allow logs to be removed from deep within a forest where machines cannot easily reach. Horse logging is slower than machine harvesting and is not used in urban areas by arborists and tree service companies. However, a team of horses might be the only way to get logs out of otherwise inaccessible areas in urban parks. Combined with sawing and other urban forestry activities, horse logging could be part of an urban park public festival. There is an association for horse-loggers, the North American Horse and Mule-Loggers Association, Inc. (http://www.pacinfo.com/~dfrench/horselogging/main.html). See http://www.uky.edu/OtherOrgs/AppalFor/draftl.html for a brief discussion of the productivity of horse logging versus mechanical skidding. See Ben Hoffman. December/January, 2000. "Working Wisely in the Woods", *Sawmill & Woodlot Management*: 4(2), 33-34, 37-38 for methods of low-impact skidding with horses.

2. National Institute for Occupational Safety and Health. "No Evidence That Back Belts Reduce Injury Seen in Landmark Study of Retail Users", press release, December 5, 2000. A large-scale study conducted by NIOSH and The Centers for Disease Control and Prevention, reported December 6, 2000 in the *Journal of the American Medical Association*, found no evidence that belts prevent injuries among retail workers. Granted, log and lumber lifting, and lifting merchandise in retail stores, are not exactly the same. However, given that logs and lumber are as heavy as, if not heavier than, the objects retail workers typically lift, a reasonable inference to draw from this study is that belts provide even less protection when lifting logs and lumber. See www.cdc.gov/niosh/belt-inj.html for more information.

3. Manual haulers and dollies specifically made for logs are available. The largest can carry a log up to 30 inches in diameter, 12 feet long, and weighing as much as 2,700 pounds (at 5 feet in width, it will not fit through 3-foot gates). However, these log-specific carriers are not likely to be available at most tool-rental stores. Local arborists may own one and might be willing to rent it. For more information, see Sherrill, Inc. listing of arboriculture, rescue, and recreational equipment at www.wtsherrill.com. Also see Bailey's Logging Supply at http://www.baileys-online.com/ for logging and related equipment.

4. Wallisky, John. June/July, 2001. "Building a Homemade Log Arch", *Sawmill & Woodlot*, No. 23: 22, 24.

5. Three better-known manufacturers are Logosol, Inc., Granberg International, and Haddon Tools. Their respective web sites are www.logosol.com/, www.granberg.com/, and www.haddontools.com/lumbermaker.html. Useful references for chainsaw cutting are Charles Self, 1994, *Woodworker's Guide to Selecting and Milling Wood*, Betterway Books, Cincinnati, OH: 72 - 74. The standard reference on chainsaw cutting, now out of print, is Will Malloff. 1982, *Chainsaw Lumbermaking*, Taunton Press, Newtown, CT. Two articles are available in *Fine Woodworking*, 1986, *Wood and How to Dry It*, Taunton Press. One is by Malloff (18 - 21) and the other by Sperber (12 - 16).

6. Four better-known manufacturers are Laimet, Lucas Mill, Mighty Mite®, Mobile Manufacturing Company. Their respective web sites are: www.laimet.com/eng/

csm.html, http://www.baileys-online.com/Mill.htm, http://home.pacifier.com/~mytmite/, and www.mobilemfg.com/ framesproducts.htm).

7. In addition to Wood-Mizer® band saw mills, other manufacturers are Timber Harvester and Timberking. Their respective web sites are: www.woodmizer.com/, www.timberharvester.com/, and http://www.timberking.com/.

8. Avery, Thomas Eugene and Harold E. Burkhart. 2002. *Forest Measurements*, McGraw-Hill, New York: 113.

9. Frank Freese. no date. *A Collection of Log Rules*, General Technical Report, U.S. Department of Agriculture, Forest Service, Forest Products Laboratory, Madison, WI: 1.

10. Avery, Thomas Eugene and Harold E. Burkhart, 117.

11. National Weather Service website, http://205.156.54.206/om/trwbro.htm.

12. National Weather Service web sites, http://205.156.54.206/om/wcm/lightning/factsheet.htm and http://205.156.54.206/om/wcm/lightning/little_known_facts.htm. If you think that the 90 percent who are injured resume their lives as usual, you should read one man's account of being struck and the consequences for his life. See Michael Utley, "My Personal Bolt of Lightning", *New York Times*, Monday, August 27, 2001: A21.

13. F. B. Malcolm. October, 2000. *A Simplified Procedure for Developing Grade Lumber From Hardwood Logs*, Research Note FPL-RN-098, U.S. Department of Agriculture, Forest Service, Forest Products Laboratory, Madison, WI. Procedures described in this publication apply to circular sawmills where the cutting plane is perpendicular to the ground. In my summary, I have, in effect, rotated the procedures 90° so that they apply to band mills that saw in a plane horizontal to the ground.

14. If the log is being sawed to meet National Hardwood Lumber Association (NHLA) standards and the first boards are expected to be the highest grades (Select or First and Seconds, designated FAS) then the minimum width of the narrowest face of the first board should be about 7 inches from one end of the board to the other end (for a minimum length of 8 feet). This exceeds the minimum 6 inch width required for a board to be graded FAS (the extra allows for shrinkage and edge trimming). If the boards are expected to be lower grade (No. 1 Common or lower) then the width can be about 4 inches since the NHLA standard allows a minimum 3 inch width. These boards can be as short as 4 feet in length. For a basic sawing guide for sawyers, see Gene Wengert. 2000. *From Woods to Woodshop: A Guide for Producing the Best Lumber*, Wood-Mizer Products, Inc., Indianapolis, IN. Of course, you can have the logs sawn to meet other requirements, such as those of a specific buyer or a specific project. If there are no other specific requirements then sawing to NHLA standards assures the broadest marketability of the lumber.

15. For more information about the *Hardwood Market Report* (weekly report), go to http://hmr.com, call (901) 767-9126, or write to P.O. Box 241325, 845 Crossover Lane, suite 103, Memphis, TN, 38124-1325. For more information about the *Weekly Hardwood Review*, go to www.hardwoodreview.com, or call (704) 543-4408, or write to P.O. Box 471307, Charlotte, NC 28247-1307.

16. National Hardwood Lumber Association. 1998. *Rules for the Measurement & Inspection of Hardwood & Cypress*. National Hardwood Lumber Association, Memphis, TN. As a way of establishing a national hardwood-grading standard, lumber buyers and manufacturers founded this organization in 1898. In addition to standard setting, the NHLA also serves as the agreed-upon arbiter in disputes between buyers and sellers over the grade(s) of a particular shipment of lumber. It also provides three-month full-time training in hardwood grading. Mark Horne, Chief Inspector for the NHLA, emphasized that grading is a professional skill acquired by formal training followed by three to five years of full-time professional experience (telephone communication, August 24, 2001). You can contact the NHLA (901 377-1818) for the names of certified and experienced graders in your area. Two other NHLA publications provide an introductory guide to hardwood grading: *An Illustrated Guide to Hardwood Lumber Grades*. 1994 (revised) and *An Introduction to Grading Hardwood Lumber*, 1994.

17. National Institute of Standards and Technology. September, 1999. *American Softwood Lumber Standard*, Voluntary Product Standard DOC PS 20-99, U.S. Department of Commerce, Gaithersburg, MD. Though softwood grading has been done since the 1920s, the separate standards were consolidated into a single uniform standard in 1970 (PS 20-70). With only minor exceptions, American Lumber Standard Committee (ALSC) standards are now used throughout the U.S. to grade softwoods and also to grade softwood imported from Canada.

18. Strength is the resistance to stress where stress is measured as the amount of pressure applied to a given area. An example would be a board under a pressure of 2,000 pounds per square inch. Wood's response to stress is strain, which is measured as the amount a board deforms (bends or is compressed), under pressure, divided by its original dimension when not under pressure. For example, the stress measure for a board 12 inches long under pressure at both ends that is shortened by .006 of an inch would be .0005 inches per inch (.006/12). Wood responds to stress in three stages. In the first stage, it deforms in constant proportion to steady increases in pressure. When the pressure is released, the wood fully recovers to its original shape or size. In the second stage, the wood passes its proportional limit. Now it deforms more

quickly under constant increases in pressure and does not fully recover after the pressure is released; that is, some strain remains impressed in the wood even after it is no longer under pressure. Permanent set is the difference between its original and its new shape or size after the pressure is released. Finally, pushed too far, the wood fails and breaks. In using wood structurally, we want to stay in the first or elastic stage where the wood flexes under a load but then would fully recover if the load were removed. Wood can be stressed four ways: by compression (being squeezed), tension (being pulled apart or stretched), shear (when a board breaks and the pieces go in different directions), and bending (when a board suspended between two points flexes or bends under a load). Wood is orthotropic, meaning that it has separate mechanical properties in three directions: along the grain, across grain, and tangent to the grain. Wood responds differently to stress depending on which of these three directions the pressure is applied. The strength of a piece of wood (its resistance to stress) of a given size depends on the species, density, moisture content, and the presence of defects such as decay, knots, shake (separation between annual growth rings), splits, wane (part of the board is either missing or, what is structurally the same thing, it is bark), or pitch pockets. With minor exceptions, only softwoods are stress-graded since only they, not hardwoods, are used in construction. The source cited in the previous footnote, DOC PS 20-99, provides the most recent grading standards for softwood lumber in the U.S. If you are thinking about using the wood you cut in construction, you might want to read the following: R. Bruce Hoadley's *Understanding Wood: A Craftsman's Guide to Wood Technology*, chapter 4 and the Forest Products Society's *Wood Handbook, Wood as an Engineering Material*, chapters 5 and 6.

19. Telephone communication with Tom Hanneman, Western Wood Products Association, August 27, 2001.

20. www.randomlengths.com/newrl.html

21. Architectural Woodwork Institute. 1999. *Architectural Woodwork Quality Standards*, 7th edition, version 1.2, Reston, VA. For more detailed information on AWI standards, go to www.awinet.org, or call (703) 733-0600, or write to 1952 Isaac Newton Square W, Reston, VA 20190.

22. There are other more specialized standards. The Wood Components Manufacturers Association (WCMA) sets its own standards for manufactured wood products. Examples are edge-glued panels, turnings, dowels, moldings, staircase parts, flooring, and cabinet parts and doors. The Association represents manufacturers of products made from hardwoods, softwoods, plywood, and engineered wood. For more specific information, go to www.woodcomponents.org. The Maple Flooring Manufacturers Association (MFMA) promotes the use of northern hard maple, yellow birch, and beech wood flooring products and provides information about flooring and sports flooring systems. MFMA's members are flooring manufacturers, installation contractors, distributors, and related product manufacturers. For more information, go to www.maplefloor.org. There is an equivalent organization for oak flooring, the National Oak Flooring Manufacturers Association NOFMA). This Association administers grading rules for hardwood flooring that covers unfinished oak, beech, birch, maple, hickory and pecan, ash, and pre-finished oak. See www.nofma.org.

23. Tom Koetter. December, 1996. "Experiences with Dimension", *Sawmill Production of Hardwood Dimension Parts: A Guide for Potential Manufacturers and Users*, U.S. Department of Agriculture, Forest Service, Northeastern Area, State and Private Forestry, St. Paul, MN: 7, 8. The following aptly sums up the point: Early in the history of his company, Tom Koetter discovered what many secondary producers already know—lumber made to National Hardwood Lumber Association (NHLA) rules was not the optimum size for making his products. Ideally, he needed 7-foot long, 13/16-inch thick lumber to make the doors his company specialized in and 9/16-inch thick lumber for base trim. Since he could not buy lumber with these dimensions, waste was generated by his use of 8-foot standard 4/4 lumber. His entry into the dimension business helped him control the volume of this waste and converted a waste product into a valuable product line. His advice is, "Do not let grade rules blind you to the real value of wood as used by the [buyer]."

24. We actually made several attempts to cut near-veneer thick pieces before we succeeded. However, we were cutting limb wood from a 500-year-old bur oak, which for us was about as challenging as it gets in thin sawing and resawing. One other hint when gluing thin pieces: spread the glue very thinly and clamp the entire surface as completely as possible (vacuum clamps might work better than mechanical clamping). Because it is so thin, the wood absorbs moisture from the glue and has a pronounced tendency to buckle as it dries. The first top we did was perfect, there were no visible seams between the wedges and the grain of each wedge mirrored the grain of the adjacent wedge. Unfortunately, even though it was thoroughly (mechanically) clamped, every wedge buckled. We finally got it right on the fifth try.

25. One cautionary note: wood with circular or wavy grain will be torn out in small but noticeable chunks (like divots from a golf course fairway) when fed into a thickness planer. The planer's blades encounter grain facing in the direction opposite to the rotation of the blades. Instead of shaving a thin layer off, the wood is torn from the surface of the board. A drum

sander will thin and smooth a board without tear-out, but at a much slower rate.

26. Hardwood and softwood consumption numbers come from James L. Howard. April, 2001. *U.S. Timber Production, Trade, Consumption, and Price Statistics 1965-1999*, Research Paper FPL-RP-595, U.S. Department of Agriculture, Forest Service, Forest Products Laboratory, Madison, WI, Table 12: 33. Numbers for pallet productions are from Jeffery John Bejune. 2001. *Wood Use Trends in the Pallet and Container Industry: 1992-1999*, unpublished M.S. thesis, Department of Wood Science and Forest Products, Virginia Polytechnic Institute and State University, Blacksburg, VA, Table 1.

27. Go to http://nwpca.com/search_state.htm.

28. Another way to compensate for cupping is to rip the cupped board into several boards of lesser widths, flatten each one in the planer, and then glue them back together. This may be a better solution than planing the entire board when cupping is pronounced (or the board is wider than what the planer will take). The amount each narrower board is cupped is less than the amount for the board as a whole. However, for me, a major reason for using flat sawn boards is the cathedral grain pattern, especially when book-matched. Cutting a wide board into narrower pieces and then gluing them back together breaks the continuity of the pattern in a noticeable way which, in part, defeats the reason for using flat sawn boards to begin with.

29. Wood-Mizer Products, Inc. *Basic Concepts Regarding Sawing and Drying Lumber*, Form 601, Indianapolis, IN: 1-3, 4.

30. For those interested in Stickley's work from the woodworker's perspective, see A. Patricia Bartinique. 1992. *Gustav Stickley: His Craft*, The Craftsman Farms Foundation, Parsippany, NJ and Joseph J. Bavaro and Thomas L. Mossman, *The Furniture of Gustav Stickley*. 1996. Linden Publishing Inc. Fresno, CA.

31. Forest Products Laboratory. 1999. *Wood Handbook, Wood as an Engineering Material*, General Technical Report, FPL-GTR-113. U.S. Department of Agriculture, Forest Service, Forest Products Laboratory, Madison, WI: 3-2 - 3-4.

32. More precisely, the wood is dried in an oven for 24 hours just above the 100°C boiling point of water. James E. Reeb, "Drying Wood", *Online Publications, University of Kentucky*, http://www.ca.uky.edu/agc/pubs/for/for55/for55.htm

33. Forest Products Laboratory. 1999. *Wood Handbook, Wood as an Engineering Material*, General Technical Report, FPL-GTR-113. U.S. Department of Agriculture, Forest Service, Forest Products Laboratory, Madison, WI: 3-8, 3-11.

34. Joseph Denig, Eugene M. Wengert, and William T. Simpson. September, 2000. *Drying Hardwood Lumber*, Gen. Tech. Rep. FPL-GTR-118. Madison, WI: 9.

35. R. Bruce Hoadley. 2000. *Understanding Wood: A Craftsman's Guide to Wood Technology*, Taunton Press, Newtown, CT: 117.

36. Joseph Denig, Eugene M. Wengert, and William T. Simpson. September, 2000. *Drying Hardwood Lumber*, Gen. Tech. Rep. FPL-GTR-118. Madison, WI: 10.

37. Air holds water in the form of water vapor (water in its gaseous state). Water vapor is measured in grains where 1 grain = 1/7000th of a pound. Absolute humidity is the number of grains of water vapor in a cubic foot of air (or, by metric measure, the number of grams in a cubic meter). The number of grains a cubic foot of air can hold depends on its temperature. As air temperature rises, so does its capacity to hold water vapor. Holding capacity roughly doubles for every rise of 18°F. For example, at 50°F, a cubic foot holds about 5 grains of water vapor while at 70°F it holds about 8 grains. Relative humidity is the amount of water vapor in a cubic foot divided by the maximum amount that cubic foot could hold at a given temperature. For example, at 50°F, a cubic foot can hold a maximum of 5 grains. If it is holding 3 grains then its relative humidity is 3 grains ÷ 5 grains x 100 = 60 percent relative humidity. It is the relationship between the amount of water vapor in the air, measured by relative humidity, and the MC of wood that determines whether wood absorbs or desorbs water.

38. For a given RH, the cell walls of wood give up water a bit easier than they can take it back. Known as the hysteresis effect, this means that the EMC of wood losing water is slightly higher than the EMC of drier wood taking in water. The reason is that water molecules are somewhat more difficult to absorb back into the cell walls than they were to desorb from the walls. The EMC of wood taking in water is 85 percent of the EMC of the wood losing water. Forest Products Laboratory. 1999. *Wood Handbook, Wood as an Engineering Material*, General Technical Report, FPL-GTR-113. U.S. Department of Agriculture, Forest Service, Forest Products Laboratory, Madison, WI: 3 - 7. This effect might be the origin of the notion, circulating as conventional wood wisdom, that once wood is kiln dried to a particular MC it stays there. This is 15 percent correct and 85 percent incorrect.

39. This is a very brief overview of kiln drying. Much has been written on this subject. If you have an interest in learning more, the following are good sources: Joseph Denig, Eugene M. Wengert, and William T. Simpson. September, 2000. *Drying Hardwood Lumber*, Gen. Tech. Rep. FPL-GTR-118. Madison, WI; William T. Simpson, (ed.). August, 1991. *Dry Kiln Operator's Manual* (revised), Agriculture Handbook 188, U.S. Department of Agriculture, Forest Service, Forest Product Laboratory, Madison, WI: and, Eugene Wengert. 1990. *Drying Oak Lumber*,

Department of Forest Ecology and Management, University of Wisconsin, Madison, WI.

40. William T. Simpson, (ed.). August, 1991. *Dry Kiln Operator's Manual* (revised), Agriculture Handbook 188, U.S. Department of Agriculture, Forest Service, Forest Product Laboratory, Madison, WI: 48 - 49. Kilns can also be grouped by how the lumber is loaded—by forklift in separate packages (package-loaded) or on kiln trucks that roll on tracks (track-loaded). Package-loaded kilns are generally smaller than track-loaded ones and are used mainly to dry hardwoods. Softwoods are dried in the larger track-loaded kilns. A distinction is also made according to heat source within the kiln—steam, heated air (referred to as direct fire), and solar radiation. Dehumidification kilns remove moisture given off by the lumber from the air inside the kiln by moving the moist air across cold coils. The moisture condenses into water and, as it does, it gives up energy, which is used to heat the dehumidified air that is then returned to the kiln. This recycling dries the lumber. Powered by electricity, these kilns are considered to be more efficient than those that use steam. The latter vent the heated moisture-laden air while the dehumidification kiln reuses its energy. Other methods use infrared radiation, electromagnetic energy, and vacuum chambers.

41. For a small kiln built around a household dehumidifier from Sears, see Editorial Staff. June, 2002. "Dry Your Own Wood", *American Woodworker*: 42 - 55. Plans for a homemade solar kiln are described in Peter J. Stephano. June, 1994. "Wood Magazine Builds a Solar Kiln", *Wood Magazine*: 44 - 46. For basic advice on owning and operating a first kiln, see Eugene M. Wengert, "Dry lumber kiln—buying your first", http://www.woodweb.com/knowledge_base/Dry_lumber_kiln.html.

42. See Joseph Denig, Eugene M. Wengert, and William T. Simpson. September, 2000. *Drying Hardwood Lumber*, Gen. Tech. Rep. FPL-GTR-118. Madison, WI: 4. and William T. Simpson, John L. Tschernitz, and James J. Fuller. October, 1999. *Air Drying of Lumber*, General Technical Report, FPL-GTR-117. U.S. Department of Agriculture, Forest Service, Forest Products Laboratory, Madison, WI: 21.

43. The equilibrium moisture content (EMC) percentages by month for selected cities in the U.S. and other nations are available in William T. Simpson. August, 1998. *Equilibrium Moisture Content of Wood in Outdoor Locations in the United States and Worldwide*, Research Note, FPL-RN-0268, U.S. Department of Agriculture, Forest Service, Forest Products Laboratory, Madison, WI. Percentages for U.S. cities are in Appendix H.

44. Approximate air drying times for specific species by selected city are available in William T. Simpson and C. A. Hart. November, 2000. *Estimates of Air Drying Times for Several Hardwoods and Softwoods*, General Technical Report FPL-GTR-121, Forest Products Laboratory, Forest Service, U.S. Department of Agriculture, Madison, WI.

45. Telephone communication with Sujit Banerjee, Institute of Paper Science and Technology, Atlanta, GA, October 02, 2001. To determine whether a very large quantity of wet wood drying indoors poses a hazard, an estimate would have to be made of the total amount of methanol and formaldehyde given off. This would be compared to the maximum allowable amount to determine whether a hazard exists.

46. William W. Rice. September/October, 1996. "Seasoned Wood: What You Need to Know", *Fine Woodworking*: 71.

Chapter Four

1. Personal communication with Ann Gatch (November 6, 2001), great-granddaughter of John Newton Gatch and Kathy Gatch (November 7, 2001), great-great granddaughter. The description of this project first appeared in my article "Bur Oak", *Popular Woodworking*, September, 1999: 78, 80.

2. Under the right conditions, bur oak (Quercus macrocarpa), sometimes known as mossycup oak, can live up to 300 years and reach 80 feet in height. It usually has a short stout trunk that supports long branches and an open majestic crown. The cap that holds the acorn is elliptical in shape, up to two inches in length, and has a fringe along its edge (hence, mossycupped or burred). Sam Benvie. 2000. *The Encyclopedia of North American Trees*, Firefly Books, Ltd., Buffalo, NY: 225 - 227. As indicated in Appendix E, the heartwood is highly resistant to rot. It is often used outdoors for railroad ties, utility poles, and mine timbers. Indoor uses include flooring and veneer.

3. Telephone communication with Brian Ritter, (former) Forestry Supervisor (May 4, 2001), Biltmore Estate.

4. Telephone communication with Parker Andes, Director of Horticulture (June 19, 2002), Biltmore Estate.

5. Personal interview with Mrs. Judy O'Bannon (March 1, 2002). Information also supplied by Nancy J. Cira, Governor's Residence Director, State of Indiana.

6. Diatomaceous earth (DE) is processed diatomite, a sedimentary rock composed of the skeletal remains of diatoms, single cell marine and lake algae-like plants. Composed mainly of amorphous silica, it is used as a filtering agent, absorbent, and liquid clarifier. DE is also widely used as a non-toxic insecticide in grain storage, livestock bedding and feed, and in gardening. Instead of poisoning insects, DE sticks to and ultimately inflicts so many cuts on their exoskeletons that they lose body moisture and eventually die of dehydration. Food grade DE can be ingested without

harm by animals and humans. It should not be inhaled and should be kept out of the eyes.

7. Santa Barbara Museum of Natural History, Chumash Indian Life. See http://www.sbnature.org/chumash.

8. Stephen M. Bratkovich. October, 2001.*Utilizing Municipal Trees: Ideas from Across the Country*, NA-TP-06-01, U. S. Department of Agriculture, Forest Service, Northeastern Area, St. Paul, MN: 23 - 26.

9. Personal interview with Cindy McCall (July 26, 1999) and interviews by telephone (June 18 and 19, 2002).

10. Steven Bodzin. September/October, 2000. "From Urban Trees to Treasured Timber", *City Trees*, 36(5).

11. Lompoc's success with the CDF band mill loan project has led to the recommendation that Bakersfield could also benefit and should apply for the use of a mill. Erika Stockton. Winter/Spring, 2002. *Rescuing a Valuable Natural Resource: Milling Urban Trees*, unpublished paper, California State University, Bakersfield, CA.

12. Telephone communication with Jeffrey Melin, June 27, 2002.

13. For more information, see http://www.eastwest-wood.com/main_flash.htm.

14. Quote from John Balzar. March 31, 2002. "Money Does Grow on Trees", Los Angeles Times, M5.

15. Balzar, M5.

16. The Rainforest Alliance is an international non-profit organization whose mission is to preserve the world's tropical forests by creating economically and socially workable alternatives to deforestation. The Alliance's worldwide SmartWood^{cm} program provides certification for wood that comes from sustainably managed forests as well as a chain-of-custody certification for products manufactured from approved wood. The program was itself accredited for forest management certification in 1996 by the Forest Stewardship Council, a nongovernmental organization headquartered in Oaxaca, Mexico that establishes standards worldwide for certification programs. Begun in 1999, the SmartWood^{cm} Rediscovered Wood Program provides a seal of approval for products made from reused and salvaged wood. Participating companies agree to use salvaged wood from trees felled because they are diseased or hazardous, or to accommodate landscaping, gardening, or modest additions to existing buildings. Trees from orchards that have reached the end of their productive lives are also acceptable, especially when they are to be replaced by new trees. Unacceptable sources are woods and forests cleared for real estate development or agricultural expansion. Commercial forestry removals are not included but might be certified under separate guidelines for forests and plantations. This seal of approval has a rising market value because a growing number of buyers of wood and wood products want to know that the products they purchase have not been produced in a way that further depletes the world's forests. As environmental awareness spreads, more consumers want environment-friendly production to be one of the attributes of the goods they buy, from tuna to wood. For more information on SmartWood^{cm}, see: http://www.smartwood.org/. For more on the Rediscovered Wood Program, see: http://www.brandsystems.net/smartwood/.

17. A half dozen more are related in Stephen M. Bratkovich, *Utilizing Municipal Trees: Ideas From Across the Country*, NA-TP-06-01, U. S. Department of Agriculture, Forest Service, Northeastern Area, St. Paul, MN. Some of the same stories are also related in a technical report done for the California Department of Forestry and Fire Protection by Tim R. Plumb, Marianne M. Wolf, and John Shelly. May, 1999. *California Urban Woody Green Waste Utilization, Urban Forests Ecosystems Institute*, California Polytechnic State University, San Luis Obispo, CA: 59 - 79 (available as a pdf document at http://www.ufei.calpoly.edu/data/abstracts/urban-wood/document.html). Warren Wise's The Woodsman in Stockton, California and Don Seawater's Pacific Coast Lumber both utilize urban trees. The Peter Lang Company, just outside Santa Rosa, California, specializes in walnut lumber and burls from groves in northern California.

18. Kathleen Newton. June, 2000. "Up from the ashes", *Woodshop News*, 14(7): T29-T33 and personal interview with Dave Parmenter on July 31, 2000.

19. Stephen M. Bratkovich: 57.

20. George Hessenthaler. Fall 1993. "The Case for Salvaging 'Waste Wood'", *California Trees*, 4(4): 3.

21. Stephen M. Bratkovich: 57.

22. Serena Herr. Fall, 1993. "Harvesting Urban Forests", *California Trees*, 4(4): 2. Anonymous. August/September, 1996. "Reclaiming the Pacific Northwest: SmartWood - Rediscovered Certifies Into the Woods", *The Canopy*: 2.

Chapter Five

1. See Mark Warda. 2000. *How to Form a Nonprofit Corporation*, Sphinx Publications, Clearwater, FL or Anthony Mancuso. 2002. How to Form a Nonprofit Corporation (Fifth edition), Nolo Press, Berkeley CA.

APPENDIX A

Commercial Lumber Names with Related Tree and Botanical Names*

Hardwoods

Commercial Name For Lumber	Tree name	Botanical name
Alder, Red	Red Alder	*Alnus rubra*
Ash, Black	Black Ash	*Fraxinus nigra*
Ash, Oregon	Oregon Ash	*Fraxinus latifolia*
Ash, White	Blue ash	*Fraxinus quadrangulata*
	Green ash	*Fraxinus pennsylvanica*
	White ash	*Fraxinus americana*
Aspen (popple)	Bigtooth aspen	*Populus grandidentata*
	Quaking aspen	*Populus tremuloides*
Basswood	American basswood	*Tilia americana*
Beech	American beech	*Fagus grandifolia*
Birch	Gray birch	*Betula populifolia*
	Paper birch	*Betula papyrifera*
	River birch	*Betula nigra*
	Sweet birch	*Betula lenta*
	Yellow birch	*Betula alleghaniensis*
Box Elder	Boxelder	*Acer negundo*
Buckeye	Ohio buckeye	*Aesculus glabra*
	Yellow buckeye	*Aesculus octandria*
Butternut	Butternut	*Juglans cinerea*
Cherry	Black cherry	*Prunus serotina*
Chestnut	American chestnut	*Castanea dentata*
Cottonwood	Balsam poplar	*Populus balsamifera*
	Eastern cottonwood	*Populus deltoides*
	Black cottonwood	*Populus trichocarpa*
Cucumber	Cucumbertree	*Magnolia acuminata*
Dogwood	Flowering dogwood	*Cornus florida*
	Pacific dogwood	*Cornus nuttallii*
Elm, Rock	Cedar elm	*Ulmus crassifolia*
	Rock elm	*Ulmus thomasii*
	September elm	*Ulmus serotina*
	Winged elm	*Ulmus alata*
Elm, Soft	American elm	*Ulmus americana*
	Slippery elm	*Ulmus rubra*
Gum	Sweetgum	*Liquidambar styraciflua*
Hackberry	Hackberry	*Celtis occidentalis*
	Sugarberry	*Celtis laevigata*
Hickory	Mockernut hickory	*Carya tomentosa*
	Pignut hickory	*Carya glabra*
	Shagbark hickory	*Carya ovata*
	Shellbark hickory	*Carya lachinosa*

Commercial Name For Lumber	Tree name	Botanical name
Holly	American holly	*Ilex opaca*
Ironwood	Eastern hophornbeam	*Ostrya virginiana*
Locust	Black locust	*Robinia pseudoacacia*
	Honeylocust	*Gleditsia triacanthos*
Madrone	Pacific madrone	*Arbutus menziesii*
Magnolia	Southern magnolia	*Magnolia grandiflora*
	Sweetbay	*Magnolia virginiana*
Maple, Hard	Black maple	*Acer nigrum*
	Sugar maple	*Acer saccharum*
Maple, Oregon	Big leaf maple	*Acer macrophyllum*
Maple, soft	Red maple	*Acer rubrum*
	Silver maple	*Acer saccharinum*
Oak, Red	Black oak	*Quercus velutina*
	Blackjack oak	*Quercus marilandica*
	California black oak	*Quercus kelloggi*
	Cherrybark oak	*Quercus falcata*
	Laurel oak	*Quercus laurifolia*
	Northern pin oak	*Quercus elipsoidalis*
	Northern red oak	*Quercus rubra*
	Nuttall oak	*Quercus nuttalli*
	Pin oak	*Quercus palustris*
	Scarlet oak	*Quercus coccinea*
	Shumard oak	*Quercus shumardii*
	Southern red oak	*Quercus falcata*
	Turkey oak	*Quercus laevis*
	Willow oak	*Quercus phellos*
Oak, White	Arizona white oak	*Quercus arizonica*
	Blue oak	*Quercus douglasii*
	Bur oak	*Quercus macrocarpa*
	Valley oak	*Quercus lobata*
	Chestnut oak	*Quercus prinus*
	Chinkapin oak	*Quercus muehlenbergii*
	Emory oak	*Quercus emoryi*
	Gambel oak	*Quercus gambelii*
	Mexican blue oak	*Quercus oblongfolia*
	Overcup oak	*Quercus lyrata*
	Post oak	*Quercus stellata*
	Swamp chestnut oak	*Quercus michauxii*
	Swamp white oak	*Quercus bicolor*
	White oak	*Quercus alba*

Commercial Name For Lumber	Tree name	Botanical name
Oregon Myrtle	California laurel	*Umbellularia californica*
Osage Orange	Osage orange	*Maclura pomifera*
Pecan	Bitternut hickory	*Carya cordiformis*
	Nutmeg hickory	*Carya myristiciformis*
	Water hickory	*Carya aquatica*
	Pecan	*Carya illinoensis*
Persimmon	Common persimmon	*Diospyros virginiana*
Poplar	Yellow poplar	*Liriodendron tulipifera*
Sassafras	Sassafras	*Sassafras albidum*
Sycamore	Sycamore	*Platanus occidentalis*
Tanoak	Tanoak	*Lithocarpus densiflorus*
Tupelo	Black tupelo, blackgum	*Nyssa sylvatica*
	Ogeechee tupelo	*Nyssa ogeche*
	Water tupelo	*Nyssa aquatica*
Walnut	Black walnut	*Juglans nigra*
Willow	Black willow	*Salix nigra*
	Peachleaf willow	*Salix amygdaloides*

Softwoods

Commercial Name For Lumber	Tree name	Botanical name
Cedar		
Alaska	Yellow cedar	*Chamaecyparis nootkatensis*
Eastern Red	Eastern redcedar	*Juniperus virginiana*
Incense	Incense cedar	*Libocedrus decurrens*
Northern White	Northern whitecedar	*Thuja occidentalis*
Port Orford	Port Orford cedar	*Chamacyparis lawsonia*
Southern white	Atlantic whitecedar	*Chamaecyparis thyoides*
Western red	Western redcedar	*Thuja plicata*
Cypress		
Baldcypress	Baldcypress	*Taxodium distichum*
Pond cypress	Pond cypress	*Taxodium d. var. nutans*
Fir		
Alpine	Subalpine fir	*Abies lasiocarpa*
Balsam	Balsam fir	*Abies balsamea*
California Red	California red fir	*Abies magnifica*
Douglas	Douglas fir	*Pseudotsuga menziesii*
Fraser	Fraser fir	*Abies fraseri*
Grand	Grand fir	*Abies grandis*
Noble	Noble fir	*Abies procera*
Pacific grand	Pacific silver fir	*Abies amabalis*
White	White fir	*Abies concolor*
Hemlock		
Carolina	Carolina hemlock	*Tsuga caroliniana*
Eastern	Eastern hemlock	*Tsuga canadensis*
Mountain	Mountain hemlock	*Tsuga mertensiana*
Western	Western hemlock	*Tsuga heterophylla*
Juniper		
Western	Alligator juniper	*Juniperus deppeana*
	Rocky Mountain	*Juniperus scopulorum*
	Utah juniper	*Juniperus osteoperma*
	Western juniper	*Juniperus occidentalis*
Larch		
Western	Western larch	*Larix occidentalis*
Pine		
Bishop	Bishop pine	*Pinus muricata*
Coulter	Coulter pine	*Pinus coulteri*
Digger	Digger pine	*Pinus sabibiana*
Knobcone	Knobcone pine	*Pinus attenuate*
Idaho white	Western white	*Pinus monticola*
Jack	Jack pine	*Pinus banksiana*
Jeffrey	Jeffrey pine	*Pinus jeffreyi*
Limber	Limber pine	*Pinus flexilis*
Lodgepole	Lodgepole pine	*Pinus contorta*
Longleaf	Longleaf pine	*Pinus palustris*
	Slash pine	*Pinus elliottii*
Northern white	Eastern white	*Pinus strobus*
Norway	Red pine	*Pinus resinosa*
Pitch	Pitch pine	*Pinus rigida*
Ponderosa	Ponderosa pine	*Pinus ponderosa*
Southern Pine major	Loblolly pine	*Pinus taeda*
	Longleaf pine	*Pinus palustris*
	Shortleaf pine	*Pinus echinata*
	Slash pine	*Pinus elliottii*
Southern Pine minor	Pond pine	*Pinus serotina*
	Sand pine	*Pinus clausa*
	Spruce pine	*Pinus glabra*
	Virginia pine	*Pinus virginiana*
Southern Pine mixed	Loblolly pine	*Pinus taeda*
	Longleaf pine	*Pinus palustris*
	Shortleaf pine	*Pinus echinata*
	Slash pine	*Pinus elliottii*
	Virginia pine	*Pinus virginiana*
Radiata	Monterey pine	*Pinus radiata*

**Source: Kent A. McDonald and David E. Kretschmann. 1999. "Commercial Lumber", Wood Handbook, Wood as an Engineering Material, General Technical Report, FPL-GTR-113, U.S. Department of Agriculture, Forest Service, Forest Products Laboratory, Madison, WI: 5-5, 5-13.*

APPENDIX B

Weight of Green Logs by Species and Diameter*

Multiply log length in feet by the weight of a 1-foot section using the average log diameter.

	Weight in lbs./ cubic ft	Weight in lbs. per one-foot sections—based on average diameters													
		10"	12"	14"	16"	18"	20"	22"	24"	26"	28"	30"	32"	34"	36"
Apple	55	30	43	59	77	97	120	145	173	203	235	270	307	347	388
Ash, white	48	26	38	51	67	85	104	126	150	177	205	235	267	302	338
Basswood	42	23	33	45	59	74	92	111	132	155	180	206	235	265	297
Beech	54	29	42	58	75	95	118	142	169	199	231	265	301	340	381
Birch (paper)	50	27	39	53	70	88	109	132	157	184	214	245	279	317	353
Birch (yellow)	57	31	45	61	80	101	124	151	179	210	244	280	319	360	403
Butternut	46	25	36	49	64	81	100	121	144	170	197	226	257	290	325
Cherry (black)	45	25	35	48	63	79	98	119	141	166	192	221	251	283	318
Chestnut	55	30	43	59	77	97	120	145	173	203	235	270	307	347	388
Cottonwood	49	27	38	52	68	86	107	129	154	180	209	240	273	310	346
Elm (American)	54	29	42	58	75	95	118	142	169	199	231	265	301	340	381
Gum (black)	45	25	35	48	63	79	98	119	141	166	192	221	251	283	318
Gum (red)	50	27	39	53	70	88	109	132	157	184	214	245	279	317	353
Hackberry	50	27	39	53	70	88	109	132	157	184	214	245	279	317	353
Hickory (shagbark)	64	35	50	68	89	113	140	169	201	236	273	314	357	403	452
Honeylocust	61	33	48	65	85	108	133	161	192	225	261	299	341	385	431
Magnolia (evergrn)	59	32	46	63	82	104	129	156	185	217	252	289	329	372	417
Maple (red)	50	27	39	53	70	88	109	132	157	184	214	245	279	317	353
Maple (silver)	45	24	35	48	63	79	98	119	141	166	192	221	251	283	318
Maple (sugar)	56	31	44	60	78	99	122	148	176	206	239	275	313	353	396
Oak (black)	62	34	48	66	86	109	135	163	194	228	265	304	346	390	437
Oak (live)	76	41	60	81	106	134	166	200	238	280	324	372	424	478	536
Oak (red)	63	34	49	67	88	111	137	166	198	232	269	309	251	397	445
Oak (white)	62	34	48	66	86	109	135	163	194	228	265	304	346	390	437
Osage-orange	62	34	48	66	86	109	135	163	194	228	265	304	346	390	437
Pecan	61	33	47	65	85	108	133	161	192	225	261	299	341	385	431
Persimmon	63	34	49	67	88	111	137	166	198	232	269	309	250	397	445
Poplar, yellow	38	21	30	40	53	67	83	99	119	140	162	186	211	239	268
Sassafras	44	24	34	47	61	78	96	116	138	162	188	215	245	277	310
Sycamore	52	28	41	55	72	92	113	137	163	191	222	254	290	327	366
Walnut (black)	58	32	45	62	81	102	126	153	182	213	248	284	323	364	409
Hemlock (eastern)	50	27	39	53	70	88	109	132	157	184	214	245	279	317	353
Pine (N. white)	36	20	28	38	50	64	78	95	113	133	154	176	201	227	254
Spruce, red	34	19	27	36	47	60	74	90	106	125	145	166	189	214	239
Tamarack	47	26	37	50	65	83	102	124	147	173	200	230	262	295	331

* *A. Robert Thompson. 1955 (revised). Rope Knots and Climbing, National Park Service, Tree Preservation Bulletin No. 7, p. 19.*

APPENDIX C
Correct and Incorrect Lifting Methods

Always Lift Gradually

Source: Forest Industry Safety & Training Alliance, Inc. April, 1999. Loggers' Safety Training Guide, Rhinelander, Wisconsin, pg. 6, 7.

APPENDIX D

Certified Softwood Grading Organizations in the U.S.*

Accredited by the Board of Review

American Lumber Standard Committee
P.O. Box 210
Germantown, MD 20875-0210
(301) 540-8004

California Lumber Inspection Service (CLIS)
420 West Pine Street
Suite #10
Lodi, CA 95240
(209) 334-6956

Northeastern Lumber Manufacturers Association (NeLMA)
272 Tuttle Road
P.O. Box 87A
Cumberland Center, ME 04021
(207) 829-6901

Northern Softwood Lumber Bureau (NSLB)
272 Tuttle Road
P.O. Box 87A
Cumberland Center, ME 04021
(207) 829-6901

Pacific Lumber Inspection Bureau (PLIB)
33442 First Way South
Suite 300
Federal Way, WA 98003
(253) 835-3344

Redwood Inspection Service (RIS)
405 Enfrente Drive
Suite 200
Novato, CA 94949
(415) 382-0662

Renewable Resource Associates, Inc. (RRA)
3091 Chararral Place
Lithonia, GA 30038
(770) 482-9385

Southern Pine Inspection Bureas (SPIB)
4709 Scenic Highway
Pensacola, FL 32504
(850) 434-2611

Timber Products Inspection (TP)
P.O. Box 919
Conyers, GA 30012
(770) 922-8000

West Coast Lumber Inspection Bureau (WCLIB)
Box 23145
Portland, OR 97281-3145
(503) 639-0651

Western Wood Products Association (WWPA)
522 SW Fifth Avenue
Suite 500
Portland, OR 97204-2122
(503) 224-3930

**This list is reviewed and can change up to several times during the year. For the most current one, go to American Lumber Standard Committee's website at www.alsc.org and under the Untreated Wood Program option, select Facsimile List. There are site links for each accredited organization in the U.S. and in Canada.*

APPENDIX E
Heartwood Resistance to Decay of Selected Domestic Woods*

Resistant or very resistant	**Moderately resistant**	**Slightly or nonresistant**
Baldcypress (old growth)	Baldcypress (young growth)	Alder
Catalpa	Douglas-fir	Ashes
Cedars	Honeylocust	Aspens
Cherry (black)	Larch (western)	Basswood
Chestnut	Oak (swamp, chestnut)	Beech
Cypress (Arizona)	Pine (eastern white)	Birches
Junipers	Southern Pine (Longleaf	Buckeye
Locust (black**)	and Slash)	Butternut
Mesquite	Tamarack	Cottonwood
Mulberry (red**)		Elms
Oak:		Hackberry
Bur		Hemlocks
Chestnut		Hickories
Gambel		Magnolia
Oregon white		Maples
Post		Oak (red and black)
White		Pines (other than
Osage orange**		Longleaf, Slash and
Redwood		Eastern White)
Sassafras		Poplars
Walnut (black)		Spruces
Yew (Pacific**)		Sweetgum
		True firs (western and eastern)
		Willows
		Yellow-poplar

**Source: Forest Products Laboratory. 1999. Wood Handbook, Wood as an Engineering Material, General Technical Report, FPL-GTR-113. U.S. Department of Agriculture, Forest Service, Forest Products Laboratory, Madison, WI, p. 3-18.*

***These are exceptionally resistant to decay.*

APPENDIX F
U.S. Kiln Drying Associations

Keystone Kiln Drying Club
Gregg Lutter
Mann & Parker
335 N. Constitution
New Freedom, PA 17349-9521

New England Kiln Drying Association
William B. Smith, Exec. Director
SUNY College of Environmental
Science and Forestry
Wood Products Engineering
1 Forestry Drive
Syracuse, NY 13210-2786
Email: wbsmith@esf.edu

South Eastern Dry Kiln Club
Joe Denig
Wood Products Extension
North Carolina State University
Box 8005
Raleigh, NC 27695-8005

Great Lakes Kiln Club
Harlan Petersen
Department of Forest Products
University of Minnesota
2004 Folwell Ave
St. Paul, MN 55108
Email: hpeterse@forestry.umn.edu

Allegheny Dry Kiln Club
John Boody, Secretary
Taylor & Boody Organbuilders
Route 10 Box 58B
Staunton, VA 24401

Tennessee Valley Kiln Association
Brian Bond
Department of Forestr & Wildlife
University of Tennessee
P.O. Box 1071
Knoxville, TN 37901
Email: bbond7@utk.edu

Ohio Valley Dry Kiln Association
Carroll Fackler
University of Kentucky
Robinson Station
PO Box 982
Jackson, KY 41339-0982
Email: cfackler@ca.uka.edu

West Coast Kiln Clubs
Mike Milota
Department of Forest Products,
Oregon State
Corvallis, OR 97331-7402
Email: milotam@frl.orst.edu

Southern Dry Kiln Association
Todd Shupe
Louisiana State University
PO Box 25100
Baton Rouge LA 70894-5100
Email: tshupe@agectr.lsu.edu

NYLE: Dry Kiln Systems
Mail Address: P. O. Box 1107
Bangor, Maine, USA, 04402
Physical Address: 72 Center St.
Brewer, Maine, USA, 04412
Telephone: 207-989-4335
Toll Free U.S. + Canada 1-800-777-6953
Fax: 207-989-1101
Email: info@nyle.com

APPENDIX G

Example of Kiln Drying Prices

WILLIS LUMBER CO., INC. - CUSTOM KILN DRYING PRICES

PRICES SUBJECT TO CHANGE WITHOUT NOTICE - PHONE TOLL FREE: 1-800-FINE-LBR (346-3527) OR 740-335-2601
14 MARCH 2001
*PRICES INCLUDE $30 PER 1000 BD. FT. INITIAL HANDLING AND STICKING CHARGE

DRYING GROUP	4/4 OVER 30% M.C.	4/4 UNDER 30% M.C.	5/4 OVER 30% M.C.	5/4 UNDER 30% M.C.
MINIMUM 500 BD.FT. PER DRYING GROUP PRICES — PRICED PER 1000 BD. FT.				
APSEN, BASSWOOD BUTTERNUT, BUCKEYE, AROM.R.CEDAR, COTTONWOOD, POPLAR, W.PINE	$253	$193	$270	$204
ASH, BIRCH, CHERRY, HACKBERRY, HICKORY, H.MAPLE, S.MAPLE SASSAFRAS	$297	$215	$325	$226
BEECH, ELM, RED OAK, WHITE OAK. WALNUT SYCAMORE	$369	$264	$429	$297
MINIMUM 5000 BD.FT. PER DRYING GROUP PRICES — PRICED PER 1000 BD. FT.				
BASSWOOD, BUCKEYE BUTTERNUT, POPLAR	$226	$154	$242	$176
ASH, CHERRY HARD MAPLE, SOFT MAPLE	$270	$187	$297	$198
ELM, BEECH, RED OAK, WHITE OAK SYCAMORE, WALNUT	$341	$237	$402	$270
MINIMUM 10000 BD.FT. PER DRYING GROUP PRICES — PRICED PER 1000 BD. FT.				
BASSWOOD, BUCKEYE BUTTERNUT, POPLAR	$198	$138	$215	$149
ASH, CHERRY HARD MAPLE, SOFT MAPLE	$242	$160	$270	$171
ELM, BEECH, RED OAK, WHITE OAK SYCAMORE, WALNUT	$314	$209	$374	$242

HOW TO SEND YOUR LUMBER TO US AND AVOID EXTRA HANDLING CHARGES

A. WLC uses forklifts of 5000# capacity to unload trucks. WLC will not hand unload or hand load a vehicle. If customer hand unloads he must place 2 x 4's under the piles.
B. Lumber of the three different drying groups must be separated, if not ADD $40 M'.
C. WLC uses an automatic sticking machines so lengths must be kept in the following manner, if not add $30 per 1000' BD.FT. For lengths less than 6', add $50 per 1000 Bd.Ft.
6-7-8' TOGETHER 9-10' TOGETHER 11-12' TOGETHER 13-14' TOGETHER 15-16' TOGETHER
D. WLC will notify customer when lumber is ready for pick up. Customer will be expected to pick lumber up within seven days of notification, thereafter a charge of $10 M' per month will incur.
E. WLC charges $20.00 per M' Bd.Ft. to package dip lumber. This is recommended for all green lumber during Spring & Summer months, except for Walnut, Cherry, Maple.
F. Lumber coming in on stick and leaving on stick subtract $30/M'
G. Custom Walnut steaming is $80/M' for 8000' or more and $85/M' for under 8000'

WE OFFER SURFACE TWO SIDES & STRAIGHT LINE RIP INQUIRIES ARE WELCOMED 1-800-346-3527

APPENDIX H

Equilibrium Moisture Content (EMC) of Wood Outdoors

in Selected U.S. Cities by Month, 1997*

State	City	Jan.	Feb.	Mar.	Apr.	May	June	July	Aug.	Sept.	Oct.	Nov.	Dec.
AK	Juneau	16.5	16.0	15.1	13.9	13.6	13.9	15.1	16.5	18.1	18.0	17.7	18.1
AL	Mobile	13.8	13.1	13.3	13.3	13.4	13.3	14.2	14.4	13.9	13.0	13.7	14.0
AZ	Flagstaff	11.8	11.4	10.8	9.3	8.8	7.5	9.7	11.1	10.3	10.1	10.8	11.8
AZ	Phoenix	9.4	8.4	7.9	6.1	5.1	4.6	6.2	6.9	6.9	7.0	8.2	9.5
AR	Little Rock	13.8	13.2	12.8	13.1	13.7	13.1	13.3	13.5	13.9	13.1	13.5	13.9
CA	Fresno	16.4	14.1	12.6	10.6	9.1	8.2	7.8	8.4	9.2	10.3	13.4	16.6
CA	Los Angeles	12.2	13.0	13.8	13.8	14.4	14.8	15.0	15.1	14.5	13.8	12.4	12.1
CO	Denver	10.7	10.5	10.2	9.6	10.2	9.6	9.4	9.6	9.5	9.5	11.0	11.0
DC	Washington	11.8	11.5	11.3	11.1	11.6	11.7	11.7	12.3	12.6	12.5	12.2	12.2
FL	Miami	13.5	13.1	12.8	12.3	12.7	14.0	13.7	14.1	14.5	13.5	13.9	13.4
GA	Atlanta	13.3	12.3	12.0	11.8	12.5	13.0	13.8	14.2	13.9	13.0	12.9	13.2
HI	Honolulu	13.3	12.8	11.9	11.3	10.8	10.6	10.6	10.7	10.8	11.3	12.1	12.9
ID	Boise	15.2	13.5	11.1	10.0	9.7	9.0	7.3	7.3	8.4	10.0	13.3	15.2
IL	Chicago	14.2	13.7	13.4	12.5	12.2	12.4	12.8	13.3	13.3	12.9	14.0	14.9
IN	Indianapolis	15.1	14.6	13.8	12.8	13.0	12.8	13.9	14.5	14.2	13.7	14.8	15.7
IA	Des Moines	14.0	13.9	13.3	12.6	12.4	12.6	13.1	13.4	13.7	12.7	13.9	14.9
KS	Wichita	13.8	13.4	12.4	12.4	13.2	12.5	11.5	11.8	12.6	12.4	13.2	13.9
KY	Louisville	13.7	13.3	12.6	12.0	12.8	13.0	13.3	13.7	14.1	13.3	13.5	13.9
LA	New Orleans	14.9	14.3	14.0	14.2	14.1	14.6	15.2	15.3	14.8	14.0	14.2	15.0
ME	Portland	13.1	12.7	12.7	12.1	12.6	13.0	13.0	13.4	13.9	13.8	14.0	13.5
MA	Boston	11.8	11.6	11.9	11.7	12.2	12.1	11.9	12.5	13.1	12.8	12.6	12.2
MI	Detroit	14.7	14.1	13.5	12.6	12.3	12.3	12.6	13.3	13.7	13.5	14.4	15.1
MN	Mpls-St.Paul	13.7	13.6	13.3	12.0	11.9	12.3	12.5	13.2	13.8	13.3	14.3	14.6
MS	Jackson	15.1	14.4	13.7	13.8	14.1	13.9	14.6	14.6	14.6	14.1	14.3	14.9
MO	St. Louis	14.5	14.1	13.2	12.4	12.8	12.6	12.9	13.3	13.7	13.1	14.0	14.9
MT	Missoula	16.7	15.1	12.8	11.4	11.6	11.7	10.1	9.8	11.3	12.9	16.2	17.6
NE	Omaha	14.0	13.8	13.0	12.1	12.6	12.9	13.3	13.8	14.0	13.0	13.9	14.8
NV	Las Vegas	8.5	7.7	7.0	5.5	5.0	4.0	4.5	5.2	5.3	5.9	7.2	8.4
NV	Reno	12.3	10.7	9.7	8.8	8.8	8.2	7.7	7.9	8.4	9.4	10.9	12.3
NM	Albuquerque	10.4	9.3	8.0	6.9	6.8	6.4	8.0	8.9	8.7	8.6	9.6	10.7
NY	New York	12.2	11.9	11.5	11.0	11.5	11.8	11.8	12.4	12.6	12.3	12.5	12.3
NC	Raleigh	12.8	12.1	12.2	11.7	13.1	13.4	13.8	14.5	14.5	13.7	12.9	12.8
ND	Fargo	14.2	14.6	15.2	12.9	11.9	12.9	13.2	13.2	13.7	13.5	15.2	15.2
OH	Cleveland	14.6	14.2	13.7	12.6	12.7	12.7	12.8	13.7	13.8	13.3	13.8	14.6
OK	Oklahoma City	13.2	12.9	12.2	12.1	13.4	13.1	11.7	11.8	12.9	12.3	12.8	13.2
OR	Pendleton	15.8	14.0	11.6	10.6	9.9	9.1	7.4	7.7	8.8	11.0	14.6	16.5
OR	Portland	16.5	15.3	14.2	13.5	13.1	12.4	11.7	11.9	12.6	15.0	16.8	17.4

Equilibrium Moisture Content (EMC) of Wood Outdoors (continued)

State	City	Jan.	Feb.	Mar.	Apr.	May	June	July	Aug.	Sept.	Oct.	Nov.	Dec.
PA	Philadelphia	12.6	11.9	11.7	11.2	11.8	11.9	12.1	12.4	13.0	13.0	12.7	12.7
SC	Charleston	13.3	12.6	12.5	12.4	12.8	13.5	14.1	14.6	14.5	13.7	13.2	13.2
SD	Sioux Falls	14.2	14.6	14.2	12.9	12.6	12.8	12.6	13.3	13.6	13.0	14.6	15.3
TN	Memphis	13.8	13.1	12.4	12.2	12.7	12.8	13.0	13.1	13.2	12.5	12.9	13.6
TX	Dallas-Ft.Worth	13.6	13.1	12.9	13.2	13.9	13.0	11.6	11.7	12.9	12.8	13.1	13.5
TX	El Paso	9.6	8.2	7.0	5.8	6.1	6.3	8.3	9.1	9.3	8.8	9.0	9.8
UT	Salt Lake City	14.6	13.2	11.1	10.0	9.4	8.2	7.1	7.4	8.5	10.3	12.8	14.9
VA	Richmond	13.2	12.5	12.0	11.3	12.1	12.4	13.0	13.7	13.8	13.5	12.8	13.0
WA	Seattle-Tacoma	15.6	14.6	15.4	13.7	13.0	12.7	12.2	12.5	13.5	15.3	16.3	16.5
WI	Madison	14.5	14.3	14.1	12.8	12.5	12.8	13.4	14.4	14.9	14.1	15.2	15.7
WV	Charleston	13.7	13.0	12.1	11.4	12.5	13.3	14.1	14.3	14.0	13.6	13.0	13.5
WY	Cheyenne	10.2	10.4	10.7	10.4	10.8	10.5	9.9	9.9	9.7	9.7	10.6	10.6

EMC values were determined from the average of 30 or more years of relative humidity and temperature data available from the National Climatic Data Center of the National Oceanic and Atmospheric Administration.

**Source: Forest Products Laboratory. 1999. Wood Handbook, Wood as an Engineering Material, Gen. Tech. Rep. FPL-GTR-113. U.S. Department of Agriculture, Forest Service, Forest Products Laboratory, p. 12-4.*

REFERENCES

American National Standards Institute, Inc. 2001. *American National Standard for Arboricultural Operations- Pruning, Repairing, Maintaining, and Removing Trees, and Cutting Brush - Safety Requirements*, ANSI Z133.1-2000, New York.

Andes, Parker. Biltmore Estate. Telephone communication, June 19, 2002.

Anonymous. November, 2000. "The State of Garbage: 12th Annual BioCycle Nationwide Survey", *BioCycle*: 40 - 48.

Anonymous. August/September, 1996. "Reclaiming the Pacific Northwest: SmartWood - Rediscovered Certifies Into the Woods", *The Canopy*: 2.

Architectural Woodwork Institute. 1999. *Architectural Woodwork Quality Standards*, 7th edition, version 1.2, Reston, VA.

Ashley, Burl S. October, 2001. *Reference Handbook for Foresters*, NA-FR-15, U.S. Department of Agriculture, Forest Service, Northeastern Area, Morgantown, WV.

Avery, Thomas Eugene and Harold E. Burkhart. 2002. Forest Measurements, McGraw-Hill, New York.

Balzar, John. March 31, 2002. "Money Does Grow on Trees", *Los Angeles Times*: M5.

Bartinique, A. Patricia. 1992. *Gustav Stickley: His Craft*, The Craftsman Farms Foundation, Parsippany, NJ

Bavaro, Joseph J. and Thomas L. Mossman. 1996. *The Furniture of Gustav Stickley*. Linden Publishing Inc. Fresno, CA

Bejune, Jeffery John. 2001. *Wood Use Trends in the Pallet and Container Industry: 1992-1999*, unpublished M.S. Thesis, Department of Wood Science and Forest Products, Virginia Polytechnic Institute and State University, Blacksburg, VA

Benvie, Sam. 2000. *The Encyclopedia of North American Trees*, Firefly Books, Ltd., Buffalo, NY.

Bernard, Bruce P. (ed.). 1997. *Musculoskeletal Disorders (MSDs) and Workplace Factors: A Critical Review of Epidemiologic Evidence for Work-Related Musculoskeletal Disorder of the Neck, Upper Extremity, and Low Back*, U.S. Department of Health and Human Services, Centers for Disease Control and Prevention, National Institute for Occupational Safety and Health, Publication No. 97-141, Cincinnati, OH. Available online at http://www.cdc.gov/niosh/pdfs/97-141b.pdf

Blanche, C. E. and H. F. Carino. October, 1996. "Sawmilling Urban Waste Logs: An Income-Generating Option for Arborists", *Arborist News*: 45 - 48.

Blenk, Stephen H. August, 1994. "Working with Burl", *American Woodworker*, 39: 48 - 51.

Bois, P. J., E. M. Wengert, and R. S. Boone. March, 1982. *A Checklist for Drying Small Amounts of Lumber*, Forest Products Utilization Technical Report No. 6, U. S. Department of Agriculture, Forest Service, Washington, DC.

Bratkovich, Stephen M. October, 2001. *Utilizing Municipal Trees: Ideas from Across the Country*, NA-TP-06-01, U. S. Department of Agriculture, Forest Service, Northeastern Area, St. Paul, MN.

Brodzin, Steven. September/October, 2000. "From Urban Trees to Treasured Timber", *City Trees*, 36 (5).

Brown, William H. 1988. *The Conversion & Seasoning of Wood*, Linden Publishing, Fresno, CA.

Bureau of the Census. November, 1994. *Geographic Areas Reference Manual*, U.S. Department of Commerce, Washington, DC.: 12-1.

Bureau of Labor Statistics, U.S. Labor Department. *Monthly Labor Review*, The Editor's Desk at

http://stats.bls.gov/opud/ted/1999/jan/wk.1/art01.htm

Caldwell, Brian. August, 2001. "Program restores elms to nation's main streets", *Woodshop News*, XV (9): 6.

Cesa, Ed. U.S. Department of Agriculture, Forest Service. Telephone communication, June 13, 2001.

Chapman, Mary. 1996. "Cheating the Landfill - Wood Waste Turned to Gold", *Understory*, winter, 6(1): 1, 6.

Consumer Product Safety Commission. no date. *Product Summary Report, All Products Injury Estimates for Calendar Year 2000*, National Electronic Injury Surveillance System, National Injury Information Clearinghouse, U.S. Consumer Product Safety Commission, Washington, DC.

Cupito, Mary Carmen and Gayle Harden-Renfro. July 27, 1982. "Tree trimmer loses arm, leg", *The Cincinnati Post*: 10C.

Cutter, Bruce E. June 15, 1994. *Considerations in Drying Hardwood Lumber*, Agricultural publication G5507, School of Natural Resources, University of Missouri-Columbia. Available online at http://muextension.missouri.edu/xplor/agguides/forestry/g05507.htm

Denig, Joseph, Eugene M. Wengert, and William T. Simpson. September, 2000. *Drying Hardwood Lumber*, General Technical Report FPL-GTR-118, U.S. Department of Agriculture, Forest Service, Forest Products Laboratory, Madison, WI.

Dramm, John. USDA Forest Service. Personal communication, November 2, 2001.

Dwyer, John F., David J. Nowak, Mary Heather Noble, and Susan M. Sisinni, August, 2000. *Connecting People With Ecosystems in the 21st Century: An Assessment of Our Nation's Urban Forests*, U. S. Department of Agriculture, Forest Service, Northeastern Research Station, Syracuse, NY.

Editorial Staff. June, 2002. "Dry Your Own Wood", *American Woodworker*, no. 70: 42 - 55.

Environmental Protection Agency. September, 1999. *Characterization of Municipal Solid Waste in the United States: 1998 Update*, EPA 530-R-99-021, Washington, DC.

Erickson, Lynn. Professional log buyer. Telephone communication, June 25, 2001.

Fisher, James A. March, 1992. *Urban Tree Residue: An Assessment of Wood Residue from Tree Removal and Trimming Operation in the Seven-County Metro Area of Minnesota*, Division of Forestry, Minnesota Department of Natural Resources.

Forest Industry Safety and Training Alliance. December, 1999. *Arborists' Chain Saw Safety Training Guide*, Rhinelander, WI.

Forest Industry Safety and Training Alliance. April, 1999. *Loggers' Safety Training Guide*, Rhinelander, WI.

Forest Products Laboratory, USDA Forest Service. October, 1999. *Air Drying of Lumber, General Technical Report*, FPL-GTR-117. U.S. Department of Agriculture, Forest Service, Forest Products Laboratory, Madison, WI.

Forest Products Laboratory. 1999. *Wood Handbook, Wood as an Engineering Material*, General Technical Report, FPL-GTR-113. U.S. Department of Agriculture, Forest Service, Forest Products Laboratory, Madison, WI.

Forest Resources Association. "Manual Timber Felling Hazard Recognition", Loss Control Overviews LCO14 under *Timber Harvesting Safety*. Available online at http://www.apulpa.org/

ForestWorld. *Woods of the World* (version 2.5), CD-ROM, Middlebury, VT.

Freese, Frank. no date. *A Collection of Log Rules*, General Technical Report, U.S. Department of Agriculture, Forest Service, Forest Products Laboratory, Madison, WI

Furiga, Paul. July 27, 1982. "Man Pinned In Tree When Limb Crashes Down", *The Cincinnati Enquirer*: D1.

Gatch, Ann. Personal communication, November 6, 2001.

Gatch, Kathy, personal communication, November 7, 2001.

Gerry, Eloise. March 25, 1914. "Tyloses: Their Occurrence and Practical Significance in Some American Woods", *Journal of Agricultural Research*, Vol. 1 (6), Washington, D.C.

Gorner, Peter. August 30, 2001. "Science crafts hardier elm", *Chicago Tribune*.

Hall, Carl T. August 23, 1998. "Staying Alive High in California's White Mountains grows the oldest living creature ever found", *San Francisco Chronicle*.

Hall, Guy H. 1998. *The Management Manufacture and Marketing of California Black Oak Pacific Madrone and Tanoak*, Western Hardwood Association, Camas, WA.

Hanneman, Tom. Western Wood Products Association. Telephone communication, August 27, 2001.

Haygreen, John G., Jim L. Bowyer. 1996. *Forest Products and Wood Science* (3rd ed.), Iowa State University Press, Ames, IA.

Hessenthaler, George. Fall 1993. "The Case for Salvaging 'Waste Wood'", *California Trees*, 4(4): 1, 3.

Herr, Serena. Fall, 1993. "Harvesting Urban Forests", *California Trees*, 4(4): 1-5.

Hoadley, R. Bruce. 2000. *Understanding Wood: A Craftsman's Guide to Wood Technology*, Taunton Press.

Hoffman, Ben. December/January, 2000. "Working Wisely in the Woods", *Sawmill & Woodlot Management*: 4(2), 33-34, 37-38.

Howard, James L. April, 2001. *U.S. Timber Production, Trade, Consumption, and Price Statistics 1965-1999*, Research Paper FPL-RP-595, U.S. Department of Agriculture, Forest Service, Forest Products Laboratory, Madison, WI

Jackson, Albert, David Day, and Simon Jennings. 1996. *The Complete Manual of Woodworking*, Alfred A. Knopf, New York.

Kasindorf, Martin. 2001. "Trees take root as energy savers", http://www.usatoday.com/news/nation/2001/07/24/tree.org

Keator, Glenn. 1998. *The Life of an Oak*. Heyday Books and the California Oak Foundation. Berkeley, CA.

Kilibarda, Frank. Minnesota Hardwoods. Telephone communication, June 15, 2001.

Koetter, Tom. December, 1996. "Experiences with Dimension", *Sawmill Production of Hardwood Dimension Parts: A Guide for Potential Manufacturers and Users, U.S. Department of Agriculture*, Forest Service, Northeastern Area, State and Private Forestry, St. Paul, MN.

Kuhns, Michael and Richard Straight. July, 1987. "Home Drying Lumber", *Working With Wood*, Cooperative Extension, Institute of Agriculture and Natural Resources, University of Nebraska-Lincoln, NB. Available online at http://www.inar.unl.edu/pubs/Forestry/g60.htm

Kuser, John (ed.). 2000. *Handbook of Urban and Community Forestry in the Northeast*, Kluwer Academic/Plenum Publishers, New York.

Leininger, Theodor, David L. Schmoldt, and Frank H. Turner. June 18, 2001. "Using Ultrasound to Detect Defects in Trees: Current Knowledge and Future Needs", *First International Precision Forestry Symposium*, University of Washington, Seattle, WA.

Lempicki, Edward. New Jersey Forestry Services. Telephone communication, June 14, 2001.

Lempicki, Edward, Ed Cesa, and J. Howard Knots. June, 1994. *Recycling Municipal Trees: A Guide to Marketing Sawlogs from Street Tree Removals in Municipalities*, U. S. Department of Agriculture, Forest Service, Washington, DC.

Lempicki, Edward, Ed Cesa. 2000. "Recycling Urban Tree Removals" in Kuser, John, ed., *Handbook of Urban and Community Forestry in the Northeast*, Kluwer Academic/Plenum Publishers, New York.

Ludivig, Dennis. Urban forester, St. Cloud, Minnesota. Telephone communication, June 21, 2001.

Malcolm, F. B. October, 2000. *A Simplified Procedure for Developing Grade Lumber From Hardwood Logs*, Research Note FPL-RN-098, U. S. Department of Agriculture, Forest Service, Forest Products Laboratory, Madison, WI.

Malloff, Will. 1986. "Chainsaw Lumbermaking", *Wood and How to Dry It*, editors of Fine Woodworking, The Taunton Press, Newtown, CN: 18 - 21.

Mancuso, Anthony. 2002. *How to Form a Nonprofit Corporation* (fifth edition), Nolo Press, Berkeley CA.

McLean, Arthur. April 22, 2002. "Urban logging raises concerns", Fayetteville Online, *Fayetteville Observer*, http://www.fayettevillenc.com//obj_stories/2002/apr/n22cut.shtml

McCall, Cindy. Personal communication, July 26, 1999. Telephone communication, June 18 and 19, 2002.

McPherson, E. Gregory and James R. Simpson. January, 1999. *Carbon Dioxide Reduction Through Urban Forestry: Guidelines for Professional and Volunteer Tree Planters*, Gen. Tech. Rep. PSW-GTR-171, U. S. Department of Agriculture, Forest Service, Pacific Southwest Research Station, Albany, CA

McPherson, E. Gregory. July, 1998. *Structure and Sustainability of Sacramento's Urban Forest*, 24(4): 174 - 190.

McPherson, E. Gregory. March, 1998. "From Nature to Nurture: The History of Sacramento's Urban Forest", *Journal of Arboriculture*, 24(2): 72 - 88.

McPherson, E. G., D. J. Nowak, and R. A. Rowntree (eds.). *Chicago's urban forest ecosystem; results of the Chicago urban forest climate project*, Gen. Tech. Rep. NE-186. Radnor, PA, U.S. Department of Agriculture, Forest Service, Northeastern Forest Experiment Station: 3 - 18, 140 - 164.

Melin, Jeffrey. West Coast Arborists, Inc. Telephone Communication, June 27, 2002.

Minnesota Department of Natural Resources. June, 1994. *Final Report: Urban Tree Utilization Project*, St. Paul, MN.

National Institute for Occupational Safety and Health. *No Evidence That Back Belts Reduce Injury Seen in Landmark Study of Retail Users*, press release, December 5, 2000.

National Institute of Standards and Technology. September 1, 1999. *American Softwood Lumber Standard*, DOC PS 20-99, U.S. Department of Commerce, Washington, DC.

Nakashima, George. 1988. *The Soul of a Tree: A Woodworker's Reflections*, Kodansha International, Tokyo.

National Hardwood Lumber Association.1998. *Rules for the Measurement & Inspection of Hardwood & Cypress*, National Hardwood Lumber Association, Memphis, TN.

National Hardwood Lumber Association. 1994 (revised). *An Illustrated Guide to Hardwood Lumber Grades*, Hardwood Lumber Association, Memphis, TN.

National Hardwood Lumber Association. 1994. *An Introduction to Grading Hardwood Lumber*, 1994. National Hardwood Lumber Association, Memphis, TN.

National Institute of Standards and Technology. September, 1999. *American Softwood Lumber Standard*, Voluntary Product Standard DOC PS 20-99, U.S. Department of Commerce, Gaithersburg, MD.

Newton, Kathleen. June, 2000. "Up from the ashes", *Woodshop News*, 14(7): T29-33.

Nowak, David J. USDA Forest Service. Telephone communication, July 30, 2001.

Nowak, David J., Mary Heather Noble, Susan M. Sisinni, and John F. Dwyer. March, 2001. "Assessing the US Urban Forest Resource", *Journal of Forestry*, 99(3): 37 - 42.

Nowak, David J. and John F. Dwyer. 2000. "Understanding the Benefits and Costs of Urban Forest Ecosystems", *Handbook of Urban and Community Forestry in the Northeast*, Kluwer Academic/Plenum Publishers, New York.

Nowak, David J. September 11 - 15, 1999. "Impact of Urban Forest Management on Air Pollution and Greenhouse Gases", *Proceedings of the Society of American Foresters, 1999 National Convention*, Portland, OR: 143 -148.

Nowak, David J. 1991. *Urban forest development and structure: analysis of Oakland, California*, Ph.D. Dissertation, University of California, Berkeley.

Occupational Safety and Health Service. February, 2001. *A Guide to Safety in Tree Felling and Cross Cutting*, Department of Labour, Wellington, New Zealand. Available online at http://www.osh.dol.govt.nz/order/catalogue/pdf/treefell.pdf).

O'Bannon, Judy. Personal communication, March 1, 2002.

Oldar, Eric. California Department of Forestry and Fire Protection. Telephone communication, June 25, 2001.

Pastoret, James. October 1, 1993. *Air Seasoning (Drying) of Wood*, School of Forestry, Fisheries and Wildlife, University of Missouri-Columbia, Columbia, MO. Available online at http://muextension.missouri.edu/xplor/agguides/forestry/g05507.htm

Phillips, Lynn. Editor, *City Trees*. Telephone communication, July 3, 2001.

Plotnik, Arthur. 2000. *The Urban Tree Book: An Uncommon Field Guide for City and Town*, Three Rivers Press, New York.

Plumb, Tim R., Marianne M. Wolf, and John Shelly. May, 1999. *California Urban Woody Green Waste Utilization*, Urban Forests Ecosystems Institute, California Polytechnic State University, San Luis Obispo, CA.

Reeb, James E. no date. "Drying Wood", *Online Publications, University of Kentucky*, http://www.ca.uky.edu/agc/pubs/for/for55/for55.htm

Rice, Robert W., Jeffrey L. Howe, R. Sidney Boone, John L. Tschernitz. May, 1994. *Kiln Drying Lumber in the United States: A Survey of Volume, Species, Kiln Capacity, Equipment, and Procedures, 1992-1993*, General Technical Report, FPL-GTR-81, U. S. Department of Agriculture, Forest Service, Forest Products Laboratory, Madison, WI.

Rice, William W. September/October, 1996. "Seasoned Wood: What You Need to Know", *Fine Woodworking*: 68 - 71.

Rietz, Raymond C. September, 1978. *Storage of Lumber*, Agriculture Handbook No. 531, U. S. Department of Agriculture, Forest Service, Forest Products Laboratory, Madison, WI.

Ritter, Brian, formerly of Biltmore Estate, May 4, 2002.

Royce, Rebecca. December, 1995. "Program aims to turn 'waste' trees into source of profitable sawlogs", *Woodshop News*: T37 - T39.

Schmidt, Udo. July, 2000. "How Wood Dries", *Woodshop News*: T32 - T33, T40.

Self, Charles. 1994. *Woodworker's Guide to Selecting and Milling Wood*, Betterway Books, Cincinnati, OH

Short, Ryan L. U.S. Bureau of the Census. Telephone communication, August 7, 2001.

Simpson, William T. August, 1998. *Equilibrium Moisture Content of Wood in Outdoor Locations in the United States and Worldwide*, Research Note FPL-RN-0268, U. S. Department of Agriculture, Forest Service, Forest Products Laboratory, Madison, WI.

Simpson, William T. (ed.). August, 1991. *Dry Kiln Operator's Manual* (revised), Agriculture Handbook 188, U. S. Department of Agriculture, Forest Service, Forest Products Laboratory, Madison, WI.

Simpson, William T. and C. A. Hart. November, 2000. *Estimates of Air Drying Times for Several Hardwoods and Softwoods*, General Technical Report FPL-GTR-121, U. S. Department of Agriculture, Forest Service, Forest Products Laboratory, Madison, WI.

Smiley, E. Thomas, Bruce R. Fraedrich, and Peter H. Fengler. 2000. "Hazard Tree Inspection, Evaluation, and Management", John Kuser (ed.), *Handbook of Urban and Community Forestry in the Northeast*, Kluwer Academic/Plenum Publishers, New York.

Smith-Fiola, Deborah. 2000. "Integrated Pest Management", Kuser, John (ed.). 2000, *Handbook of Urban and Community Forestry in the Northeast*, Kluwer Academic/Plenum Publishers, New York: 261 - 286.

Sperber, Robert. 1986. "Chain-saw Lumbermaking", *Wood and How to Dry It*, editors of *Fine Woodworking*, The Taunton Press, Newtown, CT: 12 - 16.

Stephano, Peter J. June, 1994. "Wood Magazine builds a Solar Kiln", *Wood Magazine*, no. 94: 44 - 46.

Stockton, Erika. Winter/Spring, 2002. *Rescuing a Valuable Natural Resource: Milling Urban Trees*, unpublished paper, California State University, Bakersfield, CA.

Summit, Joshua and E. Gregory McPherson. March, 1998. "Residential Tree Planting and Care: A Study of Attitudes and Behavior in Sacramento, California", *Journal of Arboriculture*, 24(2): 89 - 111.

Sygnatur, Eric F. Winter, 1998. "Logging is Perilous Work", *Compensation and Working Conditions*, Bureau of Labor Statistics, Washington, DC: 3 - 9.

Tschantz, Barbara A., Paul L. Sacamano. 1994. *Municipal Tree Management in the United States*, Davey Resource Group and Communication Research Associates.

USDA Forest Service, Southern Region and Southern Research Station, Southern Group of State Foresters. 2002, in progress. "Benefits and Costs of the Urban Forest", *Urban Forestry: A Manual for the State Forestry Agencies in the Southern Region*, some chapters are available at http://www.urbanforestrysouth.org/pubs/ufmanual/index.htm.

Utley, Michael. Monday, August 27, 2001. "My Personal Bolt of Lightning", *New York Times*: A21.

Vieth, Phil. Formerly with the Minnesota Department of Natural Resources. Telephone communication, June 20, 2001.

Vondriska, George. August, 2000. "Bandsaw Resawing", *American Woodworker*, no. 81: 46 - 51.

Wallisky, John. June/July, 2001. "Building a Homemade Log Arch", *Sawmill & Woodlot*, No. 23: 22, 24.

Warda, Mark. 2000. *How to Form a Nonprofit Corporation*, Sphinx Publications, Clearwater, FL

Wengert, Eugene M. "Dry lumber kiln—buying your first", http://www.woodweb.com/knowledge_base/Dry_lumber_kiln.html.

Wengert, Gene. October/November, 2001. "Back to Drying School", *Sawmill and Woodlot Management*, no. 25: 16 - 20.

Wengert, Gene. 2000. *From Woods to Woodshop: A Guide for Producing the Best Lumber*, Wood-Mizer Products, Inc., Indianapolis, IN.

Wengert, Gene. August, 1990. *Drying Oak Lumber*, Department of Forestry, University of Wisconsin-Madison, WI.

Whittier, Jack, Denise Ruse, Scott Haase. 1994. *Final Report Urban Tree Residues*

Wilson, Brayton F. 1984. *The Growing Tree* (revised edition), University of Massachusetts Press, Amherst, MA.

Wood-Mizer Products, Inc., *Basic Concepts Regarding Sawing and Drying Lumber*, Form 601, Indianapolis, IN

Wood, Virginia Steele. 1981. *Live Oaking Southern Timber for Tall Ships*, Naval Institute Press, Annapolis, MD

Xu, Zicai, Theodor D. Leininger, Andy W. C. Lee, Frank H. Tainter. 2001. "Chemical Properties Associated with Bacterial Wetwood in Red Oaks", *Wood and Fiber Science*, 33(1): 76 - 83.

Xu, Zicai, Theodor D. Leininger, Andy W. C. Lee, Frank H. Tainter. "Physical, Mechanical, and Drying Properties Associated with Bacterial Wetwood in Red Oaks", *Forest Products Journal*, 51(3): 79 - 84.

Zhu, Zhiliang. January, 1994. *Forest Density Mapping in the Lower 48 States: A Regression Procedure*, Research Paper SO-280, U.S. Department of Agriculture, Forest Service, Southern Forest Experiment Station, New Orleans, LA.

Index

M

N

O

P

U

V

W